STRUCTURAL ANALYSIS SYSTEMS

Software — Hardware
Capability — Compatibility — Applications

A. NIKU-LARI
*Director, Institute for Industrial Technology Transfer
24 Rue des Mimosas, Gournay s/Marne
F93460 France*

Volume 5

FINITE, BOUNDARY ELEMENT
&
EXPERT SYSTEMS
IN STRUCTURAL ANALYSIS

*Proceedings of the SAS World Conference
Paris, 28–30 October 1986*

PERGAMON PRESS
OXFORD · NEW YORK · BEIJING · FRANKFURT
SÃO PAULO · SYDNEY · TOKYO · TORONTO

U.K.	Pergamon Press, Headington Hill Hall, Oxford OX3 0BW, England
U.S.A.	Pergamon Press, Maxwell House, Fairview Park, Elmsford, New York 10523, U.S.A.
PEOPLE'S REPUBLIC OF CHINA	Pergamon Press, Qianmen Hotel, Beijing, People's Republic of China
FEDERAL REPUBLIC OF GERMANY	Pergamon Press, Hammerweg 6, D-6242 Kronberg, Federal Republic of Germany
BRAZIL	Pergamon Editora, Rua Eça de Queiros, 346, CEP 04011, São Paulo, Brazil
AUSTRALIA	Pergamon Press Australia, P.O. Box 544, Potts Point, N.S.W. 2011, Australia
JAPAN	Pergamon Press, 8th Floor, Matsuoka Central Building, 1-7-1 Nishishinjuku, Shinjuku-ku, Tokyo 160, Japan
CANADA	Pergamon Press Canada, Suite 104, 150 Consumers Road, Willowdale, Ontario M2J 1P9, Canada

Copyright © 1986 Pergamon Books Ltd.

All Rights Reserved. No part of this publication may be reproduced, stored in a retrieval system or transmitted in any form or by any means: electronic, electrostatic, magnetic tape, mechanical, photocopying, recording or otherwise, without permission in writing from the publishers.

First edition 1986

British Library Cataloguing in Publication Data
Structural analysis systems : software,
hardware, capability, compatibility,
applications.
Vol. 5
1. Structures, Theory of—Data processing
I. Niku-Lari, A.
624.1'71'0285   TA647
ISBN 0-08-034919-6

Cover drawing: Centrifugal pump casing.
Manufacturer: C.C.M.-Sulzer, France
Software used: CA.ST.OR

In order to make this volume available as economically and as rapidly as possible the authors' typescripts have been reproduced in their original forms. This method unfortunately has its typographical limitations but it is hoped that they in no way distract the reader.

Printed in Great Britain by Vine & Gorfin Ltd, Exmouth

INTERNATIONAL EDITORIAL ADVISORY COMMITTEE

Dr. T. ANDERSSON, *Sweden*
Prof. J. H. ARGYRIS, *Federal Republic of Germany*
Prof. K. J. BATHE, *USA*
Prof. T. BELYTSCHKO, *USA*
Dr. M. BERNADOU, *France*
Prof. B. A. BILBY, *UK*
Prof. R. D. COOK, *USA*
Dr. T. FUTAGAMI, *Japan*
Dr. S. K. GHOSH, *UK*
Dr. L. ILIE, *France*
Dr. L. IMRE, *Hungary*
Dr. J. C. LACHAT, *France*
Prof. H. LIEBOWITZ, *USA*
Dr. A. NIKU-LARI, *France [Editor]*
Ing. J. MACKERLE, *Sweden*
Ing. W. M. MAIR, *UK*
Dr. G. A. MILIAN, *Mexico*
Dr. D. NARDINI, *Yugoslavia*
Dr. I. PACZELT, *Hungary*
Prof. G. SANDER, *Belgium*
Prof. R. P. SHAW, *USA*
Prof. M. TANAKA, *Japan*
Prof. W. N. WENDLAND, *Federal Republic of Germany*
Prof. G. YOUZHONG, *People's Republic of China*

PREFACE

Artificial intelligence is the study of ideas which enable computers to do the things that make people seem intelligent [1]. But then, what is human intelligence? It is surely the ability to reason, to acquire and to apply knowledge to manipulate and communicate ideas, which the computer can not yet do.

The central goal of artificial intelligence is to make computers more useful and to understand the principles which make intelligence possible.

In modern industry, engineers make use of new and powerful tools such as structural analysis systems integrated in the engineering CAD/CAM environment. These new technologies help to solve complex problems in much shorter time. However, in the case of a bad definition of the goals much human and computer time may be wasted.

Expert systems using principles of AI can help users to choose the right method (boundary elements, finite elements, pre-stored results etc) in the right place (substructuring) at the right time (precision needed, hardware available) [2].

We hope that this book will contribute to the transfer of knowledge between research and industry in this important discipline.

This fifth volume of the SAS-international guidebook series contains papers on expert systems (Chapter 1), finite and boundary element methods (Chapter 2) and software (Chapter 3), presented at the SAS-World Conference, 28-30 October 1986 in Paris. The conference was sponsored by the French Ministry of Research and Education and was organized by IITT-international* in co-operation with AS & I**.

[1] Winston P H Artificial Intelligence. Addison-Wesley.
[2] Fouet J M Artificial intelligence in structural analysis, this volume

* IITT-International, 24 rue des Mimosas, F-93460 Gournay-sur-Marne, France
** AS & I, péripole 132, 44 rue Roger Salengro, F-94126 Fontenay sous bois, France

Papers on other main topics of the SAS-World Conference, CAD/CAM and industrial applications of structural analysis systems, are published in SAS volume four.

I would like to thank all authors and the members of the International Scientific and Advisory Committee for their commitment to the present volume.

Dr. A. KIKU-LARI
Editor

EDITORIAL

Structural analysis aims to construct numerical models which represent the best behaviour of the actual engineering material and component. These models are used in research for better understanding of experimental results. In industry the structural analysis models allow both the optimization of design and the prediction of failure.

The structural analysis (SA) is therefore a multidisciplinary problem which demands knowledge of several scientific and industrial disciplines such as, engineering sciences, mechanical or civil engineering, informatics, applied mathematics, computer sciences, etc.

International competition gives to industry the necessary impulse to optimise the design of parts and structures.

The engineer should save material and energy and use new and lighter materials such as composites. No longer is one allowed to over-design parts for "security reasons", and new international criteria have to be considered.

Industry needs to design sophisticated parts working in very special environments, in space, in the human body, in the sea, etc.

The compatibility of the structural analysis systems with modern micro-computers allow small and medium size companies to make use of these new technologies. New super computers help to find rapid solutions to complex industrial design problems.

The evolution of interactive graphics allow the full integration of structural analysis programs in a computer aided design and manufacturing environment. Expert systems, application of artificial intelligence and computer-aided decision making bring new developments in this field.

Structural analysis systems existing in the world market are powerful bridges between research and industry. They bring theory in direct physical contact with the industrial application.

The SA technology is in a rapid evolution. More and more new computers and powerful software appear in the market and the industry faces a new problem - that of selecting the optimum structural analysis system.

The choice of a structural analysis software is an important decision which can often exercise a significant influence over the successful development of a research, manufacture, or design project. Depending on the engineering problem and hardware available, a good choice of the computer program can be both cost and time effective. This new international guidebooks series aims to provide the engineer with the most up-to-date information about structural analysis systems currently available in the world market, and their capabilities.

Editorial

Published under the guidance of a distinguished scientific committee whose
members are internationally recognised specialists of finite or boundary
element methods, the series should be considered as an essential practical
reference tool for the modern engineer involved in such areas as structural,
mechanical, civil, nuclear, aeronautical and design engineering, computer
science and software development.

Each volume gives detailed information about a wide range of selected software
packages describing their purpose, capabilities and limitations and provides
several practical examples of industrial applications, often supported by case
studies. It also gives to the user the necessary information about
postprocessor capabilities, computer-aided design integration and software
compatibility with the most commonly used computers.

The guidebooks are industry-oriented and should prove indispensible in helping
potential users to select the soft and hardware most suited to their needs.
Each volume commences with a program description in tabular form, rapidly
directing readers to the program most likely to solve their industrial
problems, and concludes with a case-study index.

Main areas covered in the series:

- Finite and boundary element programs
- Finite difference and other methods
- Computer graphics
- Artificial intelligence and expert systems
- Computer-aided decision making in engineering
- Computer-aided design and manufacturing (CAD/CAM)
- Integration of structural analysis and expert systems in
  engineering CAD/CAM environment
- Hard and software selection
- Micro-computer applications in engineering
- New development in structural analysis software and interactive graphics
- Industrial case study
  . Mechanical engineering
  . Aeronautics and nuclear
  . Biomechanics
  . New materials (composites, plastics, etc)
  . Civil engineering (offshore, seismic, earthquake, etc).

Authors wishing to submit a paper under one of the above headings for possible
publication in future volumes are invited to submit their manuscript for
editorial consideration of the international scientific committee to the
address below.

Dr A Niku-Lari, Director
Institute for Industrial Technology Transfer
I.I.T.T.-international
24 Rue des Mimosas
93460 Gournay-sur-Marne
FRANCE
TEL: (1) 43.05.17.19

CONTENTS

CHAPTER 1: ARTIFICIAL INTELLIGENCE AND EXPERT SYSTEMS

I-1 ARTIFICIAL INTELLIGENCE - OVERVIEW

Artificial Intelligence in Structural Analysis   3
       J. M. Fouet

I-2 EXPERT SYSTEMS

Improved Elements for Computer Generated Design of
Dynamically Excited Structures Based on Expert Systems
and Quasi Optimisation Routines   11
       H. J. Baier

Mechanical Components Behaviour Analysis Aided by
Expert System   27
       B. Lalanne, A. Chaudouet & A. Bonnefoy

Selection and Evaluation of Structural Optimisation Strategies
by Means of Expert Systems   39
       D. Hartmann

Expert Systems in Mechanical Engineering Demonstrated by
the Selection of Flexible Couplings   57
       A. Spielvogel, W. Platt & Ch. Troeder

CHAPTER 2: FINITE AND BOUNDARY ELEMENT METHOD

An Approach to the Quality Assessment of Finite Element Meshes   69
       T. K. Hellen

A Thermal Shell Finite Element for the Thermomechanical
Analysis of Thin Shells   81
       J. L. Blanchard & E. Carnoy

On a Method of Structural Analysis 93
 Guo Youzhong

Finite Element Analysis of Fatigue Crack Growth Process 107
 A. Bia & G. Pluvinage

The Boundary Element Method as a CAD Tool 123
 C. A. Brebbia

CHAPTER 3: SOME ENGINEERING SOFTWARE

BEASY: A Boundary Element System for Structural
Analysis (abstract) 167
 C. A. Brebbia

KYOKAI: A 2/D Potential and Elasticity Problem Solver
on Super Mini 169
 K. Onishi

MODULEF: A Library of Computer Procedures for Finite
Element Analysis (abstract) 173
 M. Bernadou, P. L. George, P. Laug & M. Vidrascu

MEF/MOSAIC: Linear, Nonlinear Finite Element Code with
Interactive Graphic Display (abstract) 175
 J. F. Cochet

NELIN3: Computer Program for Nonlinear Analysis of
Space Frame and Cable Structures 177
 J. Ozbolt

NISA: The Performance of NISA, a General Purpose Finite
Element Program for Static, Dynamic and Heat Transfer
(Linear and Nonlinear) Analysis 185
 K. S. Kothawala

SYSTUS: An Industrial Structural Analysis System at
the Service of Industry 195
 D. Halbronn and J. C. Chaumont

Subject Index 203

Chapter 1
ARTIFICIAL INTELLIGENCE
AND
EXPERT SYSTEMS

ARTIFICIAL INTELLIGENCE IN STRUCTURAL ANALYSIS

Jean-Marc Fouet

*ENS de Cachan, Université Paris 6, Greco "Calcul des Structures",
61 Avenue du Président Wilson, 94230 Cachan, France*

ABSTRACT

The cost of Structural Analysis is still too high to allow its introduction at all levels of Design, inside optimization loops. Large scale non linear problems are still hardly solvable. On the other hand, experts can be found who solve problems much faster than computers do, despite the slowliness of nervous influx. We discuss the methods Artificial Intelligence offers to make these ends meet, give a few examples of existing applications, and suggest reasonable fields of research.

KEY WORDS

Artificial Intelligence. Structural Analysis. Knowledge Representation. Heuristics. Interfaces. Supervisors.

INTRODUCTION

Ten years ago, "Artificial Intelligence" was a dirty word. Nowadays, with so-called Expert Systems spreading all over the technological world, no one cares anymore whether A.I. is a science, a technique or an art : it's a tool everybody wants to use. But this might backfire once engineers realize they have been sold, at expensive cost, nothing but fancy programming languages. The best way to avoid such a situation is certainly to understand the methods, to clearly define the classes of problems they can be applied to, and to be aware of their limitations. Based on experience in CAD and FE applications, the present paper tries to look at A.I. from the Structural Analysis point of view, or the other way round.

STUCTURAL ANALYSIS SEEN FROM THE A.I. POINT OF VIEW.

Haber (1982) writes : "In practice, input costs represent about 80 per cent of the total cost of analysis". Smith (1986) shows examples of "systematically misleading outputs" and "mismodeling". In other words, three major causes for waste can be exposed :
- time spent by humans preparing and analysing computations ;
- time spent by computers ;
- time spent by both on computations that are wrong anyway.

The first point leads to research on interfaces, helping people to define their problem and retrieve its solution. The third one imposes constraints on these interfaces, all the more since the increasing role of computers in engineering has a tendency to atrophy the physical intuitions of Engineers. We will come back later to these points, but let us emphasize the second one.

Problems with tens of thousands of degrees of freedom are not uncommon. Dynamical, non linear problems of that size are scarce. Or, to be more precise, scarce are the engineers who dare ask for the computer time necessary to solve that kind of problem. Computer Science offers two answers to that question : on the one hand, "Try and develop faster algorithms, optimize them, throw in some heuristics" ; on the other hand, "Buy a super computer". Artificial Intelligence has a third answer : "Think". If someone can say : "Yes, there is a quicker way, but it's so complicated I don't even know where to start modelizing it", then A.I. can probably help. This, of course, is not in contradiction with software (Bussy, 1985) and hardware (Norrie, 1982) efforts, which it tries to integrate, as we will see after a few definitions.

KEYWORDS AND METHODS

Symbolic Processing

The french word "Ordinateur" makes no assumption on the way a "Computer" should be used. Computers can handle any type of symbol, and perform any type of operation over them : they are not afraid to deal with thousands of symbols, and they make no mistake. Then why compute the minimum of a function, iteratively and with an admitted error, when one could differentiate it, solve the equation, and thus get the <u>exact</u> form of the solution, once and for all possible values of the parameters ?

Declarative vs. Procedural

Stupid people, and computers, are given orders and obey them. Intelligent people are given goals, and fulfil them. They are told what to do, not <u>how</u> to do it. If anything happens that was not foreseen by the programmer, the former fail: the latter succeed without even noticing, or start discussing the goal. Such behaviour, of course, requires two difficult features :
- enough background knowledge to be able to find a path towards the goal ;
- knowledge expressed in a descriptive form, not a prescriptive form.

These "educational" problems will be discussed at length throughout the rest of this paper.

Modularity

An old survey made by the D.o.D. in 1975 showed that it cost $60 to write a correct program statement, no matter what programming language was used, and $4000 to add a new statement to an existing program. The reason for that enormous difference lies within the _structure_ of a program : almost any statement has an influence on almost any other. Sometimes that influence was obvious at the time of programming, but is later forgotten. In most cases, that influence is a side-effect, unsuspected until a modification reveals it.

In order to avoid such costs, particularly if one knows from the start that one's system will need to be often modified, methodologies have been put up, based on fierce modularity. But these programming languages still include loops and alternatives, thus enabling programmers to build complex, therefore fragile, structures. Artificial Intelligence has turned instead to another representation : Logic.

Logic

Logic meets the two requirements mentioned above : it is modular and declarative. Production Rules ("if A then B") can be seen as molecules in a gas, whereas programming statements are molecules in a cristal. If one adds, subtracts or modifies such "chunks" of knowledge, the system will behave slightly differently, but not collapse as a program does. Production Rules show a relationship between two or more facts ("if A is true then B is true") : they do not compel anyone to do anything. Moreover, they can be interpretated in different ways : deducing that B is true from the fact that A is, deducing that A is false if B is, focusing on A if you want to prove B, etc.

Incremental Development

Experts are people who know things no one else knows. This knowledge was not found in books or at school: it was gained on-the-job, as problems arose and were solved. Therefore, when you ask such an expert to tell you what he knows, how he solves a problem, you get only vague answers. McDermott (1981) showed that the only way to secure knowledge from these people is to subject a possible solution to their criticism, translate their remarks into modifications of the system, and iterate until the expert acknowledges he could not have done any better. This works because experts, although they _don't know what they know_, and are unable to tell what should be done, are very good at explaining why a solution is wrong.

This process of incrementally developing systems imposes two constraints :
- programming has to be banished, for the reasons mentioned above ;
- the expert should be available, motivated, and patient : one of the major pitfalls about Expert Systems is the eagerness to transfer to a computer in two weeks what an engineer learnt in five years.

Interpretation vs. Compilation

A Compiler is a program that transforms a program, written in a language understandable by humans, into a program executable by the machine. During that translation, optimization occurs. An Interpreter is a program that transforms one statement of the given program, executes it, and then moves on to the next statement. It is of course much slower, since it re-analyzes the same statement each time it comes to it, and cannot have a global optimizing point of view. On the other hand, it is sometimes helpful to humans to see, step by step, how their program is being executed.

One of the arguments often quoted in favor of expert systems is that their behaviour is "logical" and that they can explain how and why they produce their results. This is true of <u>interpretated</u> expert systems. But it is also true that they are awfully slow. Fast systems can only be achieved by <u>compilation</u> ; in that case, the external, logical, declarative, modular representation used as input is transformed into an internal, procedural, highly structured representation.

When one learns a new task, for instance driving, one is given rules : "Before turning, check in your rear-view mirror", "If you come to a crossing, watch the traffic lights", etc. During the learning stage, one constantly goes through a check-list of these rules : "Am I in such and such situation ? No. Am I in that situation ? Yes. Then what I have to do is...". Later, as one gains habits, one reacts to situations, unconsciously and much faster : one doesn't interpret knowledge anymore, one follows compiled procedures. Once this has been achieved, it is very difficult to explain how one gets through the Place de l'Etoile at 6 pm in less than a minute : we have no decompiler. Neither have efficient systems.

Having thus emphasized a few points which we consider important, let us go back to Structural Analysis.

HOW CAN A.I. HELP

Leaving aside the related fields of CAD (Reynier, 1984) and Optimization (Arora, 1986), let us focus on Structural Analysis proper. Where do we feel we waste time ?

Input

As soon as one mentions input to Finite Elements, one thinks of mesh generators. Other aspects of the problem should not be underestimated. Having been written before the widespreading of conversational processing, by people who were not Computer Scientists, most Finite Element Codes offer rather crude interfaces, batch oriented. Therefore, puzzled beginners merely copy examples from the manual, changing as little parameters as possible. Later, this turns into recipes that one follows without exactly knowing why, and most of all without being aware there might be a faster way of solving their problem. There is need for expert systems having good knowledge about FE in general, a given code in particular, and users : those systems should be given the problem and infer the way to formulate the input to the code ; in other words we need program synthesis (Barstow, 1979).

Mesh generation was probably the first field of Structural Analysis to be investigated with A.I. methods (Carnet, 1978). Typical A.I. tendency, here, has been to gradually specialize, from the general purpose (Carnet) through one class of mechanical part (forged axisymetrical parts, Trau, 1985), to one part only (rocket tanks, Desaleux, 1986). One of the important ideas behind these systems is that experts have their own, highly efficient representation of the problems they are solving, representation which is not geometrical. It is <u>matter</u> they mesh, matter which will be manufactured and subject to efforts. That matter is there to fulfil functions. Each part has a name, and a history. This is what the expert handles, not a mere drawing, especially if he is dealing with a draft-scheme.

Computations

A.I. can help people choose the right method (beam theory, boundary elements, finite elements, pre-stored results...) in the right place (substructuring) at the right time (precision needed, available computers). It could pilot these computations, shifting in real time from one algorithm to another, adjusting increments, stopping a diverging process. The idea is that, by modelizing one's problem, one deliberatly hides information from the computer : blind algorithms then go down these dark alleys, spending time to yield results which, the following morning, will be cast away in disgust. If a supervisor could follow these algorithms, regularly checking their progression against physical intuition, much time would be gained. Alas, I have yet no reference I can quote on this matter, but teams are at work, with the GRECO "Calcul des Structures" and elsewhere; the heaviest part of the job is to interface existing codes and their messy data structures.

Output

The "following morning" mentioned above is an understatement. It often takes days to cross-check a computation; and once an error has been found, it takes days again to trace it back to the input.

The first problem, finding the error, is interesting to A.I. in more than one aspect : people are facing a Pattern Recognition problem very much like the one computers are facing when trying to understand a photography : they are overwhelmed by enormous amounts of data, usually numerical, and they don't even know what to look for. This task is new to engineers, who did not have that problem when they used a slide-rule. They are handicapped by the fact that humans, like other animals, have transferred, during their childhood, the pattern recognition processes from their brain to their eyes. One direction of research, then, is to reconstruct visual images from numerical data ; this involves Data Analysis (Malvache, 1986) and Art (Cohen, 1984).

Debugging is an open question. At first sight, one may wonder how people use knowledge to find an error, and did not use that knowledge to prevent the error in the first place.

PITFALLS

"Conclusion : we solved all the problems but one; we therefore contemplate building an Expert System". This can more or less be found at the end of a growing number of articles. In order to avoid desillusions, we would like to remind the reader of a few principles.

Availability of Experts

If no one knows how to solve a problem, or if those who know are not prepared to spend as much time with the computer as they would have to spend with an apprentice, there will be no expert system. Remember also that, before introducing expert knowledge, you have to introduce common knowledge (maths for example) into your system.

Logics

"If you have a problem, and you solve it, then you have no problem". This looks like a Production Rule. But monotonic logic will not be able to handle such a rule, because it contains a contradiction (if A is true then A is false). "If the length is under 50 mm, then add 5mm to the length" cannot be handled either. If you need such statements, you have to turn to non monotonic logics, and that's where real trouble begins. What do you make of all the results that derived from the fact that the length was less than 50mm ?
"If there is x such that x is a triangle and the co-ordinates of one vertex of x are less than zero, then..." is a rule in first order logic. It can be evaluated from left to right (checking for all triangles whether they have a the kind of vertex we want) or from right to left (checking for all negative coordinates whether they belong to the vertex of a triangle). Depending on the size of both sets, one way may prove a thousand times faster than the other. Should you keep that in mind while writing the rule ? Besides, is it reasonable that the system should go through these evaluations, trying to find an x, when all you're interested in is the mere fact that there is such an x, no matter its value ?

Extra-logics

In most Structural Analysis applications, one needs computations : "greater than", "add", etc ; one also needs actions: "read", "mesh", etc. These can be embedded in "if/then" structures to take the form of rules. But they are not Production Rules, they are not Logic: one cannot reason about them, without knowing which numerical values their input parameters will take at the time of execution: what, for instance, should be done if, while evaluating "if x/y > 5", a divide-by-zero interrupt occurs ?

Some processes usually associated with intelligence are not handled by the brain : eyes and ears play a major part in Pattern Recognition, transferring only high-level information to the brain. Trying to modelize with Logic what goes on inside those highly complex wired peripherals is, in my opinion, a waste of time. Selling Expert Systems for that purpose is, in my opinion anyway, pure swindle.

Simplicity and Complexity

Writing a program which solves the equation of the second degree is absurd : it would take less time to find the solution by hand than it takes to write the program. Programs are worth the trouble if and only if they _iterate_. Writing an expert system that gives ready-made answers to a finite number of questions is also absurd : it would take less time to use a classical Data Base, or even a printed catalogue. Expert systems are worth the trouble if and only if they _infer_.

Incremental development, as described previously, essentially complies with a dichotomic process : the expert, analyzing why he does not agree with the system, falls upon the rule :
        if A and B then C.
"I was right to give you this rule, which applies in most cases ; but here, we have one of the exceptions". So the rule is split into :
        if A and B and not X then C   (general)
        if A and B and X then D   (particular)

Experience shows that, above a certain critical mass of about a thousand rules, this is not as simple anymore. In fact, that critical mass is more or less the number of rules the expert remembers having written : once the system is so big that the expert forgets whether he has already written, or how he wrote, certain notions, then maintenance problems arise. Work is being done in A.I. on this problem, around intelligent rule-editors and interfaces that negociate, but no significant result is yet available.

Moreover, if the reason for the failure is a bad choice of representation, expert system or not, you have to throw it away and write a new one.

CONCLUSION

Structural Analysis cannot be seen anymore as the last, verifying stage, of Design : it has to become the main tool of Optimal Design. This implies that it is easily put to use, reliable, and cheap. A.I. offers methods and tools to ease the transfer of know-how, short-cuts and reasoning power from experts to computers. But it promises no miracle.

REFERENCES

Arora, J., and G. Baenziger (1986). Uses of artificial intelligence in design optimization. Computer Methods in Applied Mechanics and Engineering, vol. 104, March.

Barstow, D.R. (1979). Knowledge based program construction. Programming Language Series, The Computer Science Library, New York.

Bussy, P., J.Y. Cognard and P. Rougee (1985). On new algorithms in plasticity and viscoplasticity. Proc. 3ème Colloque Tendances Actuelles en calcul de Structures. Pluralis, Paris.

Carnet, J. (1978). Une methode heuristique de maillage dans le plan pour la mise en oeuvre des éléments finis. Thesis, University PARIS 6.

Cohen, H. (1984). The first artificial intelligence coloring book. William Kaufmann, Los Altos Ca.

McDermott, J. (1981). R1 : a rule-based configurer of computer systems, Proc. of Expert Systems 81, London, December.

Desaleux, T. and J-M Fouet (1986). Expert systems for automatic meshing. Proc of International Conference on Reliability of Methods for Engineering Analysis. Swansea, July.

Haber, R., and J. Abel (1982). Discrete transfinite mappings for the description and meshing of three-dimensional surfaces using interactive computer graphics, Int. Journal for Numerical Methods in Engineering, vol. 18, January.

Malvache, P. (1986). ADONIA, un système expert d'aide à la conduite d'une analyse de données. Proc. Sixièmes Journées Internationales Systèmes Experts, Avignon, April.

Norrie, D.H., and C.I.W. Norrie (1982). Large scale computation : architecture and program structure for super computers and special purpose machines. Fourth International Symposium on Large Engineering Systems, June 9-11, The University of Calgary.

Reynier, M., and J-M Fouet (1984). Automated design of crankcases : the CARTER system. Computer Aided Design, vol. 16, November.

Smith, G. (1986). The dangers of CAD. Mechanical Engineering, vol. 108, February.

Trau, P. (1985). Un système de maillage tridimensionnel automatique. Thesis, University PARIS 6.

IMPROVED ELEMENTS FOR COMPUTER GENERATED DESIGN OF DYNAMICALLY EXCITED STRUCTURES BASED ON EXPERT SYSTEMS AND QUASI-OPTIMIZATION ROUTINES

Horst J. Baier

Dornier System, 7990 Friedrichshafen, FRG

ABSTRACT

Two elements for improving computer aided design of dynamically excited structures are discussed. The first is an expert system to be applied for specification of quasi-static design loads which are to be used for preliminary design and the second are efficient redesign and quasi-optimization routines which can be applied once a mathematical model is established. For the expert system the essentials from the side of the domain field as well as user requirements are discussed. Preliminary experiences in constructing and applying such a system are reported. For the quasi-optimization routines different approaches which basically lead to linear or quadratic optimization problems in combination with efficient reanalysis methods are discussed with some demonstration examples.

1. INTRODUCTION

Computer aided development of structures is becoming increasingly popular due to continuously improved CAD-packages with decreasing relative costs on the one side and increasing accuracy and efficiency in methods and software for performance prediction on the other.

But the design and analysis procedures in many cases are running in sequence with only loose formal coupling e.g. by common data bases, as the best. Efforts for synthesizing these into one common integrated design and analysis tool are under way for more than ten years. Perhaps the most formal synthesis can be carried out by mathematical programming or structural optimization approaches combining design, analysis, optimization and data management tools. Though different promising developments in this area have been made, they still can be considered more research oriented without really being in the "production phase". Reasons for this are among others the relatively low flexibility of the software in covering different complex design situations and the relatively high computational effort envolved in solving structural optimization problems especially due to the relatively high number of required reanalyses. This is especially true in the case of dynamically excited structures where a synthesized approach is limited by two major drawbacks

- in the starting phase of the structural development preliminary loads as arising from the structures dynamic response have to be specified where then the first design and analytical model is to be based upon. These preliminary loads usually are given as discretized (static) acceleration fields specified for relevant regions of the structure and its equipment. In order not to be too conservative in the beginning on the one side or to optimistic on the other, these specification are to be made from experienced experts;

- once the analytical model (usually a finite element model) based on the preliminary design is established, mathematical optimization can be applied for improving this design. But this is often limited due to the relatively high computational effort, because of the iterations necessary to achieve an optimum. In addition, minor changes of the requirements during the development process can make a strong optimum obsolete.

In order to mitigate these drawbacks, an approach based on an expert system to specify reasonable starting loads and then fast quasi-optimization routines for design improvement is selected with spacecraft structures being the application in mind. The load defining expert system consists of a data base comprising loads of different structural components obtained under previous projects, interpolation strategies, simple mechanical models and heuristic rules to combine these elements. Via a user interface such data as configuration, masses, estimated damping etc. are to be specified and the load estimates with confidence levels and some background information are given. In a further step, a more detailed analytical model is established and design improvements are carried out by quasi-optimization algorithms. This means that it is not necessarily looked for a strong optimum but good or close optimal designs for the simplification that usually not more than one structural reanalysis is required. This is achieved in principle by formulations leading to linear or quadratic optimization problems and/or approximate but simplified analyses.

2. AN EXPERT SYSTEM FOR PRELIMINARY LOAD ESTIMATION

2.1 Discussion of general background

The response of dynamically excited structures causes dynamic loads on the structure itself and especially onto the mounted equipment. Apart from the excitation, which can be harmonic, transient or random, the dynamic loads very much depend also on the dynamic properties, i.e. eigendynamics and damping, of the structure itself. So in the early beginning of the development phase one is confronted with the dilemma of specifying dynamic loads for a structure which still has to be designed, apart from the fact that a mathematical model for more sound load estimation is not yet available. In order to derive a good preliminary design where the preliminary analytical model also has to be based upon, the load specifications have to be reasonable. This means that on the one side they should envelop all loads determined later by detailed analysis but on the other should not be too conservative. The design and analysis loop is outlined in figure 2.1.

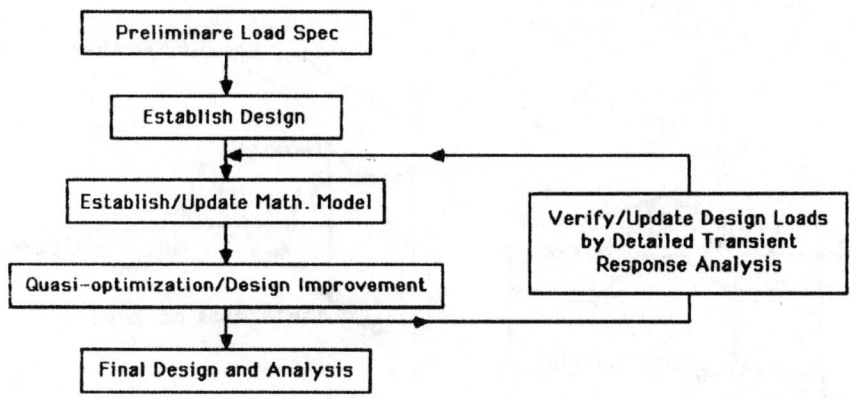

Fig. 2.1: Load specification and design loop

These preliminary design loads are to be specified by experts having good experience in the relevant type of structures, excitation etc. and a good background in structural dynamics. They combine the different areas of experience and knowledge for an "educated guess", and this is felt to be a typical area to investigate the benefits of applying an expert system. Basically such a system should then enable a structural engineering to specify such loads without necessarily having deep relevant experience and knowledge.

2.2 Requirements and elements for a dynamic load expertsystem from an engineering point of view

The expert system should enable an interested structural engineer to specify reasonable dynamic design loads. The requirements and necessary elements of such a system will be derived by answering the following questions

- what are the necessary data to be input by the user into the expert system?
- what are the necessary data be output from the expert system?
- how should the user interface look like?
- what are the necessary elements of the expert system in order to achieve the basic goals?

In order to be spcific, the questions will be discussed with the application in spacecraft dynamic loads specification. As outlined is figure 2.2, such a structure is excited by the launcher with transient and random (plus acoustic) loads.

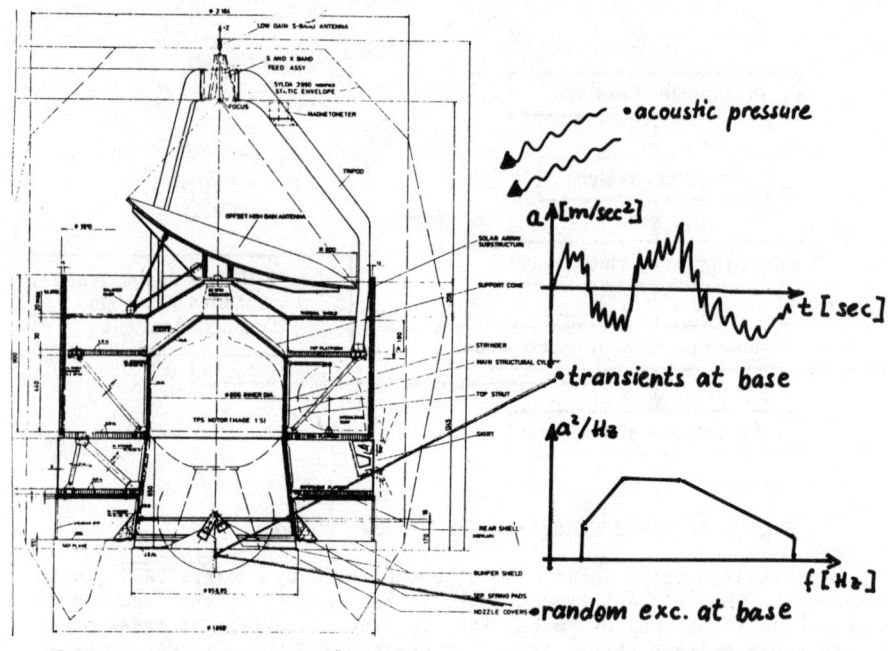

Fig. 2.2: A spacecraft structure and its excitation

The input data required are structural configuration and global dimensions, total mass and mass distribution, estimated eigenfrequency spectra, damping properties and type of launcher. The minimum information to be obtained from the expert system are design loads for overall and component design, estimated low frequency and high frequency content, if possible statistical informations such as probability of exceedance of the given loads as well as recommendations a safety factors, susceptibility to fatigue, recommendations for testing and verification etc. The elements of the expert system then will be data and minimum load requirements from launcher authority, a data bank from different projects, enveloping and interpolation functions such as the Bamford curve (see figure 2.3) as well as analytical data as derived from simplified models as given in figure 2.4. Also typical spectra from excitation and responses from previous prejects are helpful. Especially the latter elements of the expert system require it to have a window to one of the standard scientific computer languages. Correlation between the current structure and the information stored in the expert system is established via the user's input data and a kind of inference machine.

2.3 A view from the concept of expert systems

An expert system is a computer program that performs a task normally done by an expert, and in so doing it has captured heuristic knowledge. Based on 'expert rules' it avoids blind search, reasons by manipulating symbols, grasps fundamental principles of the covered domain and is producing expla-

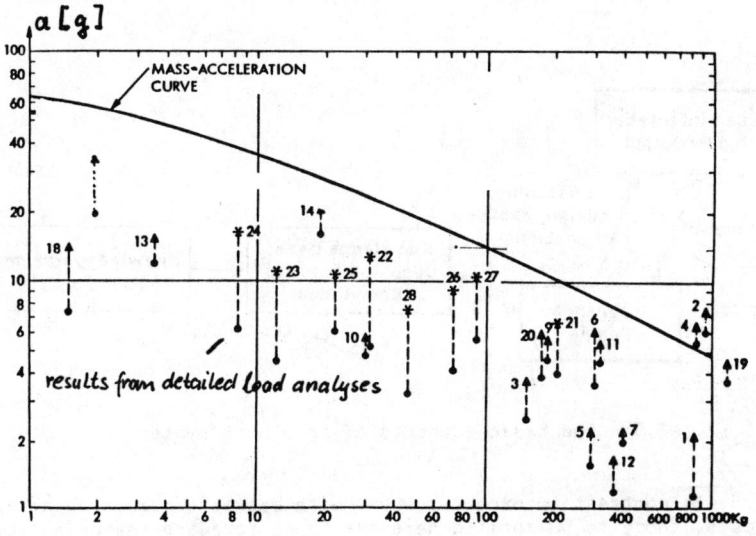

Fig. 2.3: The Bamford curve as a simple load enveloping function

nations. It combines in its knowledge base deep knowledge (so to say the basic physics) with heuristic surface knowledge coming from experience in solving a lot of problems in this domain. Its basic structure is outlined in

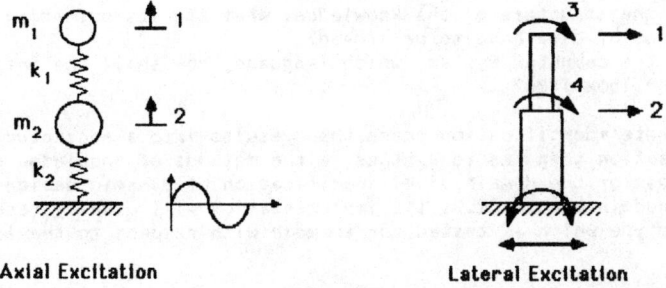

Axial Excitation   Lateral Excitation

Fig. 2-4: Simple mechanical models for quick response analysis

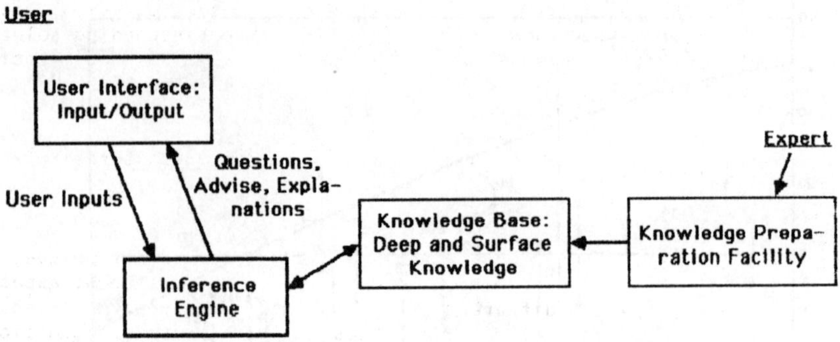

```
           User
            │
   ┌────────────────┐
   │ User Interface:│
   │  Input/Output  │
   └────────────────┘
     │    ▲                                              Expert
User │    │ Questions,                                     │
Inputs│   │ Advise, Expla-                                 ▼
     ▼    │ nations        ┌────────────────┐    ┌──────────────────┐
   ┌──────────┐            │ Knowledge Base:│◄───│ Knowledge Prepa- │
   │Inference │───────────►│ Deep and Surface│    │ ration Facility  │
   │ Engine   │            │   Knowledge    │    └──────────────────┘
   └──────────┘            └────────────────┘
```

Fi.g 2.5: The basic elements of an expert system

figure 2.5. In contrast to expert systems with backward reasoning as applied in diagnosis, those to be applied here are to be forewardreasoning for deriving a solution.

The basic question to be adressed in the following chapter now is how to construct such a system using a synthesis between the expert system's and engineering approach of the previous chapter.

2.4 Constructing the expert system

Construction of the expert system runs according the steps given as follows. In the identification phase basically questions related to a thorough definition and specification of goals and requirements are to be answered. Such typical questions are

- what's the goal of the system and its limits
- what's the structure of the knowledge, what are its essential parts?
- what type of data have to be stored?
- what's the computer system, which language, how shall the interfaces to the user look like?

A more concrete identification phase then results into a conceptualisation. The formalisation then has to synthesize the methods of knowledge engineering with those of the domain, i.e. specification of dynamic design loads, as it is outlined in chapter 2.2. Its implementation will usually result first into a prototype which is tested for example with respect to the following questions

- are the results realistic and close to those given from experts? (if not, check why)
- are the results consistent
- is the system userfriendly?

3. MATHEMATICAL AND QUASI-OPTIMIZATION OF DYNAMICALLY EXCITED STRUCTURES

In this chaper, the formulation of the design problem as a classical mathematical optimization problem is discussed first with corresponding solution algorithms addressed only briefly. Then formulations as quasi-optimization problems and solution methods are given.

3.1 Mathematical optimization of dynamically excited structures

3.1.1 Problem statements

The mathematical statement of the optimal design problem is not unique, since the design problem iteself may be formulated under different aspects and for one design problem different alternative formulations can exist. So in the following different problem statements are given with a discussion of their main properties and advantages/disadvantages.

The basic statement is

<u>Statement A</u>

$$\text{minimize } f = \text{weight } (x_1, x_2 \ldots x_n) \quad (3\text{-}1a)$$

sucht that

$$1 - u_i(X, t_i^*)/u_{ial} \geq 0 \quad , \quad i \in I \quad (3\text{-}1b)$$

$$1 - \ddot{u}_j(X, t_j^*)/\ddot{u}_{jal} \geq 0 \quad , \quad j \in J \quad (3\text{-}1c)$$

$$\omega_l^2(X)/\omega_{lal}^2 - 1 \geq 0 \quad , \quad l \in L \quad (3\text{-}1d)$$

$$1 - \sigma_k(X, t_k^*)/\sigma_{kal} \geq 0 \quad , \quad k \in K \quad (3\text{-}1e)$$

$$x_{ml} \leq x_m \leq x_{mu} \quad , \quad m = 1, \ldots n \quad (3\text{-}1f)$$

$$s_p(u_i, \ddot{u}_j, \omega_l, \sigma_k, X) = 0 \quad , \quad p = 1, \ldots \quad (3\text{-}1g)$$

with

$$t^* \text{ s.t. } \max_{t} r(X, t) = r(X, t^*) \quad .$$

The design variable vector $X^T = x_1, x_2 \ldots x_n$ has to be determined to minimize weight such that displacements u_i, accelerations $\ddot{u}_j$ and stresses σ_k do not exceed given upper bounds $u_{i\,al}$, $\ddot{u}_{j\,al}$, $\sigma_{k\,al}$, respectively. Eigenfrequencies ω_l have to exceed given lower bounds while the design variables can be lower and upper bounded by x_{ml}, x_{mu}. For given design variables the state variables can be determined from the system equations S_p which are usually provided by the finite element algorithm. Since the response data are a function of time the constraints are formulated in such a way that $t = t^*$ leads to maximum response. So far no specific assumption is made on the type of excitation, which is necessary when discussing solution strategies.

The main advantages of this basic problem statement are its straightforward formulation of the optimal design problem and the high degree of completeness and generality. On the other side, it might be difficult at least in the first development stages to define reasonable allowables especially for the accelerations. For too stringent allowables one might have to start in the infeasible region where some of the constraints are not yet satisfied; but the feasible region could also be empty, i.e. it is impossible to find a desing variable vector which satisfies all the constraints. This drawback is at least in part circumvented by the following problem statement.

Statement B

$$\text{'minimize'} \{F\} = \{u_i{}^2(X, t_i^*)\} \qquad i \in I, \; j \in J \qquad (3\text{-}2a)$$

such that

$$1 - u_i(X, t_i^*)/u_{ial} \geq 0 \;, \qquad i \in I \qquad (3\text{-}2b)$$

$$1 - \ddot{u}_j(X, t_j^*)/U_{jal} \geq 0 \;, \qquad j \in J \qquad (3\text{-}2c)$$

$$1 - \text{weight}(X)/\text{max. weight} \geq 0 \qquad (3\text{-}2d)$$

other constraints as in statement A

The square of the response data is used in (3-2a) to get rid of the sign since for example otherwise $u_i \to -\infty$ could be a solution.

Now a set of response values in critical dofs are comprised in the objective function while others with allowables easy to specify are part of the constraint function vector together with an upper bounded weight which is often part of the structural requirements. So the potential problem of empty feasible region can be circumvented. But it should be noted that the objective now consists of a vector criterion instead of a single one as in statement A. This means that

- a solution of statement B is a so called Pareto-optimal compromise between the response values in F where one arbitrary component of F can be decreased only by increasing at least one other while satisfying the constraints. It cannot be expected that for one variable vector X all components of F are at their individual minimum. (This is the reason for the quotation marks in (3-2a));

- for numerical solution a transformation to a single criterion problem with a scalar objective function is necessary. There exist different ways of doing this with different underlying assumptions and requirements. One of these approaches is a weighted linear combination of the different components:

$$f = \sum_i t_i u_i{}^2 + \sum_j w_j \ddot{u}_j{}^2 \qquad (3\text{-}3)$$

with t_i, w_j to be selected by the analyst's preference. Such transformations have to be handled with care for nonconvex problems, which one has to deal with here. As a special and often selected case of (3-3) the most critical acceleration is taken for minimization while constraining the other components.

Of course, special problems statements such as selection of optimal isolators are formally contained in those given above.

The two most relevant properties of the problems as stated above are

- they are nonlinear in X and in most cases implicit, i.e. apart from simpler examples the system response data can be obtained only numerically (usually the finite element method) thus allowing only a pointwise evaluation of the functions. This and the nonlinearity requires careful built-up of solution algorithms in order to be computationally efficient

- they are usually highly nonconvex which means a change in curvature or negative definiteness of the Hessian matrix of objective and/or constraints functions. This nonconvexity becomes obvious from the harmonic response of the one dof system as given in figure 3.1. Loosely speaking, dynamic response data in objective functions lead to 'rough' isocontours with peaks and valleys while response data in constraint functions leat to quasi disjount feasible regions especially for lower damping. The consequences of this nonconvexity are manifold. For example, the transformation (3-3) of the multi crtieria problem (3-2a) to a single criteria problem may not be complete, i.e. some parts of the solution space could be omitted. But a still more important consequence are the presence of local optimal solutions which may not be the best for the full design space. Again, this already becomes obvious from the one dof example if stiffness is the design variable to minimize the response: for $k \rightarrow \infty$ the amplification approaches 1 from above (local optimum) while for $k \rightarrow 0$ the amplification approaches zero (global optimum).

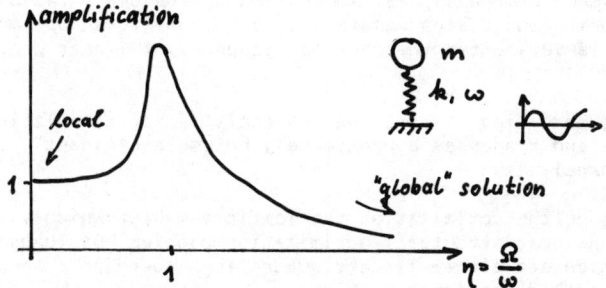

Fig. 3.1: Local and global optimum for stiffness k which minimizes the amplification ∇

3.1.2 Solution algorithms

Since the problems as stated in (3-1) and (3-2) are mathematical optimization problems, any algorithm for solving nonlinear constrained problems can be used, in principle. Two important classes adapted to the specific kind of problems are

- optimality criteria methods, where instead of approaching the optimization problem directly, optimality criteria such as the necessary Kuhn-Tucker-criterion are used which are better and better satisfied by design changes carried out by an algorithm. As shown by Khot (1985) this approach is relatively efficient if constraints are put on the eigenfrequencies only. It laks some generality for other type of constraints, especially in dynamic response.

- sequence of linear programming with the basic idea to linearize the non-linear problem by Taylor series expansion, then use a simplex algorithm to solve the linearized optimization problem and repeat this until convergence criterial are satisfied.

  This approach is relatively efficient and general and has been formulated and applied also in dynamic response problems for example by v. Nack (1981).

A more detailed discussion of these algorithms is given by Baier (1985).

3.2   Quasi.-Optimization

3.2.1 General Overview

Quasi-Optimization means considerable improvement but not necessarily a strict optimization of a structure by a computer-algorithm but with the main intention not to apply more than one or two reanalyses. This is felt to be desirable especially in the case of constrained dynamic properties since there a greater number of reanalyses is rather costly. Such quasi-optimization can be achieved by

1. applying a "standard" mathematical optimization approach for only a small number of reanalyses. In that case, the optimization algorithm should bring the design relatively fast close to an optimum, which for example is very often the case for sequence of linear programming approaches

2. applying some kind of perturbation analysis for calculating the eigendyamics (and response) approximately but more efficently than by complete reanalyses

3. formulating the optimization problem in a manner which avoids the classical nonlinear structural optimization problem but leads to formulations which are either linear, quadratic or perhaps a sequence a of smaller scaled nonlinear problems

4. combinations of the different measures

In the following, measures two and three are discussed in somewhat more detail.

Efficient reanalysis methods

There are different approximate but efficient reanalysis methods available for eigendynamics or response analysis. For eigendynamics the jth mode of the modified structure denoted by "'" for example can be approximated by a linear combination of a set of modes of the previous structure

$$\{\phi\}'_j = [\phi] \cdot \{Q\}_j \qquad (3\text{-}4)$$

with the vector {Q} to be determined. Substitution into the eigenproblem und multiplication by [φ] leads to

$$w_j^2 [\tilde{M}]' \{Q\}_j = [\tilde{K}]' \{Q\}_j \qquad (3-5)$$

with

$$[\tilde{M}]' = [\phi]^T [M]' [\phi] \qquad (3-6a)$$

$$[\tilde{K}]' = [\phi]^T [K]' [\phi] \qquad (3-6b)$$

Equation (3-5) is the eigenproblem for w_j' which is of significantly smaller order than the original one.

Alternatively, perturbation equations as discussed by Sandström (1983) which are based on [K] + [ΔK] and [φ] + [Δφ] etc. formulations lead to

$$([K] + [\Delta K])([\phi] + [\Delta\phi]) = ([\lambda] + [\Delta\lambda])([M] + [\Delta M]) \qquad (3-7)$$

which can be simplified by neglecting terms of higher order.

The response {U}', caused by a force vector {F} is obtained from the modal response {S}' via

$$\{U\}' = [\phi]' \{S\}' \qquad (3-8)$$

where {S}' results from

$$([M]'_{gen} + [K]'_{gen} + [D]'_{gen}) \{S\}' = [\phi]'^T \{F\} \qquad (3-9)$$

with [M]', [K]'$_{gen}$ being the generalized mass, stiffness and damping matrix. In the case of [M]$_{gen}$ ≡ [I], [K]$_{gen}$ = [λ] and modal damping ρ this results into the uncoupled equations

$$([I] + [\lambda]' + [\rho]) \{S\}' = [\phi]' \{F\} \qquad (3-10)$$

Now one of the eigendynamic approximate reanalysis methods can be used to determine the required matrices [λ]' and [φ]' where as a further simplification the mode shapes of the base model can be used, too.

Quasioptimization

For quasi-optimization equations (3-7), (3-10) are applied to exceed the desired frequencies or stay below maximum response values with minimum structural mass. This then leads to an optimization problem with (3-7), (3-10) as system equations. Neglection of higher than linear terms in (3-7) and eventually assuming the modes to remain unchanged result into linear optimization problems which can be solved very efficiently.

A total different approach than the previous one is to formulate the problem in a way that parameters on the right hand side of the response equations are to be determined since this again leads to linear optimization problems. This is the so called control force approach which in its basic version has been proposed by Pilkey (1976) and works as follows:

1. Sectioning of the portions of the structure to be designed and substitution by generic forces.

2. Expansion of periodic excitation and generic control forces by Fourier series.

3. Substitution into equation of motion leads to a relationship between Fourier coefficients of response and those of excitation and control forces.

4. Determination of Fourier constants of control forces such that a maximum response criterion is minimized and other constraints are satisfied. So the optimum control forces and response time functions are available.

5. "Insert" physical components at the points of control force application by determination of stiffness and damping properties of these elements such that forces caused by application of the optimal response data to these elements are as close as possible to the required ones determined under step 4.

The fourier series development of

- excitation forces

$$P(t) = \sum_i P_{c_i} \cos w_i t + \sum_i P_{s_i} \sin w_i t + P_o \qquad (3-11a)$$

- control forces (to be determined)

$$C(t) = \sum_i C_{c_i} \cos w_i t + \sum_i C_{s_i} \sin w_i t + C_o \qquad (3-11b)$$

- displacement response

$$U(t) = \sum_i U_{c_i} \cos w_i t + \sum_i U_{s_i} \sin w_i t + U_o \qquad (3-11c)$$

with

$w_i = i \cdot \pi/T$, T = half period.

Yields

$$\begin{bmatrix} -w_i^2 \underline{M} + \underline{K} & w_i \underline{D} \\ -w_i \underline{D} & -w_i^2 \underline{M} + \underline{K} \end{bmatrix} \begin{bmatrix} U_{c_i} \\ U_{s_i} \end{bmatrix} = \begin{bmatrix} \underline{V}\, C_{c_i} \\ \underline{V}\, C_s \end{bmatrix} + \begin{bmatrix} \underline{F}\, P_{c_i} \\ \underline{F}\, P_{s_i} \end{bmatrix} \qquad (3-12)$$

with $\underline{V}$ and $\underline{F}$ being time independent force coefficient matrices. What's most important is that in linear dynamics a linear relation between U_{ci}, U_{si} and the C_{ci}, C_{si} exists leading to a linear optimization problem for determination of C_{ci}, C_{si}. It should be noted that (3-12) very much simplifies for isolation problems where the eigendynamics (i.e. $\underline{K}$ and $\underline{D}$) need not be considered.

The more crucial point of this approach lies in transformation of optimal Fourier coefficients and time histories of the control force back to physical realisations as outlined in figure 3.2. Since time histories of the displacements at points of attack of the control force are known, a possible approach is to determine stiffness and damping data k_j, d_j of a discrete control element such that the resulting control force

$$c_j(t) = k_j \, (u^*_i(t) - u^*_{i-1}(t)) + d_j \, (\dot{u}^*_i - \dot{u}^*_{i-1}) \tag{3-13}$$

is as close as possible to the required time history calculated in the previous step. This leads to an approximation problem

$$\text{minimize } f = \int_0^T (c^*_j(t) - c_j(t))^P dt, \; j = 1, \ldots \tag{3-14a}$$

or since the parameters are independent

$$\text{minimize } f = \sum_j \int_0^T (c^*_j(t) - c_j(t))^P dt \tag{3-14b}$$

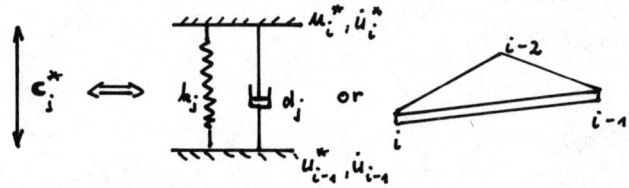

Fig. 3.2: Example to identify k_j, d_j for given c^*_j, u^*_i, $u_{i-1}*$

where the exponent p = 2 belongs t a least squares problem and the higher p the more will the larger deviations between $c^*_j(t)$ and $c_j(t)$ be minimized. Additional constraints on d_j, k_j can be easily taken into account. Consideration of general stiffness matrices is caried out in a similar way.

As the a single example consider to two massess-springs-system of figure 3.3 with a harmonic excitation applied at dof 1. The spring stiffnesses k_1 and k_2 are to be determined to minimize the response of mass 1 and the relative response between mass 1 and 2, respectively.

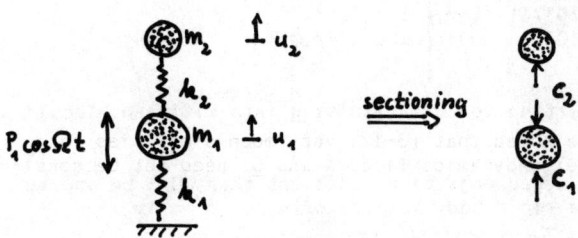

Fig. 3.3: Two masses-springs-system

Steps 1 and 2 (sectioning and expansion into Fourier series) is shown in figure 3.3. Steps 3 and 4 are the following:

- Substitution into the equations of motion to determine the Fourier constants C_{c1} and C_{c2} which minimize $\ddot{u}_1$

$$- U_{c1}\Omega^2 m_1 \cos\Omega t = (C_{c1} - C_{c2}) \cos\Omega t + P_1 \cos\Omega t \qquad (3\text{-}15a)$$

$$- U_{c2}\Omega^2 m_2 \cos\Omega t = C_{c2} \cos\Omega t \; . \qquad (3\text{-}15b)$$

As can be seen from (3-15a), C_{c1} and C_{c2} can be selected such that $\ddot{u}_1 = -U_{c1}\Omega^2 \cos\Omega t$ is zero:

$$C_{c2} = P_1 + C_{c1}$$
(equilibrium of forces)

Step 5 is the identification of k_2 such that $k_2 \cdot u_2^*(t)$ is a close as possible to $C_2^*(t)$. Since C_2 is harmonic, it can be exactly achieved with $k_2 \cdot u_2^*(t)$, since $u_2^*(t)$ is harmonic, too:

$$C_{c2} \cos\Omega t = k_2 u_2^* \cos\Omega t$$

from (3-15b):

$$u_2^* = -C_{c2}/(\Omega^2 m_2)$$

follows

$$k_2 = \Omega^2 m_2$$

i.e. the absorber solution!

If instead a "standard" mathematical optimization approach is used, then one might run into different local solutions depending on the selected starting vector for k_1 and k_2 as outlined in table 3.1

Objective fct.	Startvector/f	Finalvector/f
$f = \ddot{u}_1^2$	$10^6, 10^6/3.5$	$9 \cdot 10^5, 7.9 \cdot 10^5/10^{-4}$
	$10^4, 10^7/0.8$	$10^0, 10^9/\; 0.6$
	$10^0, 10^0/1.0$	$10^0, 10^0/1.0$
$f = (\ddot{u}_1^2 - \ddot{u}_2^2)^2$	$10^6, 10^6/56$	$10^0, 10^8/10^{-3}$
	$10^5, 10^3/11$	$10^0, 10^0/1$
	$10^4, 10^7$	$10^0, 10^8/10^{-3}$

Table 3.1: Different starting vectors resulting into different local solutions

Results of this table with larger objective functions than 10^{-3} belong to "solutions" with either free rigid body motions of m_1 or of $m_1 + m_2$.

In certain instances such as simultanuous consideration of static contraints requires the determination of control forces and the passive element approximation to be carried out in one step which leads to a quadratic optimization problem.

Taking into account static and dynamic constraints requires thorough and highly modular software design related to data management, coordination of optimization routines, determination of structural response and coordination of objective and constraints functions routines in order to achieve sufficient quality and flexibility.

REFERENCES

Dyn, Chre L, (1984). Expert systems-new approaches to computer aided engineering, Proceeding 24th AIAA/ASME Structures, Dynamics and Materials Conference, Anaheim, Cal., 1984, pp. 99-115

Chen, J., Carba, J., Salania, M., Trubert, M., A survey of loads methodologies for Shuttle Orbiter payloads, JPL Publication 80-37, June 15, 1980

Baier, Horst J. (1985), Structural optimization with constraints in dynamics, Proc. 2nd Int. Symp. on Aeroelasticity and Struct. Dyn., Aachen, W. Germ., 1985, DGLR-Report 85-02, ISBN3-9220 10-28-8

Kim, K.O., Sandström, P.E. (1983), Nonlinear inverse perturbation method in dynamic analysis, AIAA Journal, Vol. 21, Sept. 1983, pp 1310-1316

Nack, W.V. (1981), Vibrational isolating of large scale finite element models using optimization, Computers and Structures, Vol. 14, No. 1, 1981, pp. 149-152

Khot, N.S., Optimization of structures with frequency constraints, Computers and Structures, Vol. 20, No. 5, 1985, pp. 869-876

Pilkey, W., Chen Y.H., Indirect synthesis of multi degree of freedom transient systems, Journ. Optim. Theor. and Appl., Vol. 20, No. 3, 1976, pp. 331-346

MECHANICAL COMPONENTS BEHAVIOUR ANALYSIS AIDED BY EXPERT SYSTEM

B. Lalanne, A. Chaudouet and A. Bonnefoy

CETIM, 52 Avenue F. Louat, 60300, Senlis, France

ABSTRACT

Before any work, the structural engineer defines a procedure to deal with the problem. When he builds up the strategy, he is faced up to a lot of choices. These choices are important and difficult. Important, because an ill evaluation of the problem can involve an analysis which results are not efficient to conclude and can affect greatly engineer and C.P.U costs. Difficult, because the set of possible analyses is wide, application domains are numerous, and parameters are ill defined. All these reasons lead CETIM to develop a prototype expert system whose task is to help defining a strategy of analysis. The first part of this paper deals with cognitive psychology in the expert work. A reasoning typology with different hierarchical abstraction space of used knowledge is proposed. Few words are said about modelisation and representation of knowledge.

The second part elicits the inference engine. Two levels are distinguished :

- The upper level produces a sequence of suspended problems
- The lower level solves each of the precedents

For the feasibility study the knowledge base was restricted to discrete structures under static behaviour.
In the third part, the main expert knowledge, the classification between concepts, the different kind of rules and an illustrative consultation are exposed.
In the conclusion the possibilities, the limits of such a system, the involved difficulties and the extensions are presented.

KEYWORDS

Structural analysis, expert systems

INTRODUCTION

Needs of Intelligent Systems in Design and in Structural Analysis

Computer Aided Design has been contributing to rationalization of drawing in design office, to the generation of numerical commands machining data or geometrical data for scientific codes. In the recent years C.A.D has evolved toward systems which integrate the whole cycle of design process - integrated systems - to the detriment of universal ones made for a specific task (such as in mechanic, electronic and architecture) (Meziane, 1986). Presently integrated systems are limited to algorithmic problems. However, many engineering design activities including synthesis, evaluation, modelling and the creative aspects are ill-structured and require heuristic solutions based on judgment and experience (Rehak, 1985).

Knowledge based programming techniques enable to do so, but yet a unique expert system for design involves an explosion of the production rules number 1200 for a prototype in transmission shaft design (Reyner, 1984). To avoid this problem, different modules are currently separated - eventually expert ones - (Moreau, 1985 ; Rehak, 1985) such as :
- shape elaboration
- structural analysis
- cost analysis
- planification

Structural Analysis Process

In the structural analysis process two successive levels of refinment may be distinguished :
- the strategy elaboration or preliminary analysis
- the analysis itself

As a matter of fact, before any work the structural engineer defines a procedure to deal with the problem. This procedure must enhance the study progress, special difficulties and a whole vision of the problem. That's the decision step. During the analysis itself, the different strategy points are specified, decisions are taken at a more refined level of detail. Should a model be chosen in preliminary analysis as finite element then, the mesh must be layed down during the analysis process by itself. The strategy is important and difficult. Important, because an ill evaluation of the problem can involve an analysis which results are not efficient to conclude and can affect greatly engineer and C.P.U costs. Difficult, because the set of possible analyses is wide, application domains are numerous, and parameters are ill defined.

Goals

Help to the decision in preliminary structural analysis has to be furnished to the expert user. Regarding the domain complexity, we don't hope to compete with human experts. With satisfactory performances our system may be very useful in selecting the best solution and not an acceptable one and in forcing the expert to hold more rigorous reasonment.
Furthermore, the sytem has to be able to explicit its own reasoning. This is important because the user is searching help from the program and because he will take decisions following its advice (Buchanan, 1982).

EXPERT APPROACH AND REPRESENTATION OF KNOWLEDGE

Expert Approach

The consultation process itself proceeds in steps. Steps are conceptual entities useful in describing the chain of events occuring during the consultation. Many tasks may be executed during a single consultation step. Two main steps may be identified : the problem formulation and its resolution. The first one is the description and the identification of the problem. The description corresponds to the first sight of the design office plan. That's mean to classify the problem and to display the structural characteristics. So abstract concepts are instanciated (global parameters, actions, geometry) and then sorted out. At the identification stage - or diagnosis one - the ruin process (failure mode) must be determined (Hervé, 1983) : excessive distorsion progressive deformation, crack initiation...

Once the problem has been set, the expert has two ways to solve it. The first one is to use one of the multiple specifications or design code. These standards enclose an implicit knowledge of analysis activity and their use needs expertise in the selection of the design code, in the selection of the appropriate part of a code and in its interpretation. The second one is to make by himself, his own model of the structural behaviour in order to apply mechanical theories.

Knowledge Representation

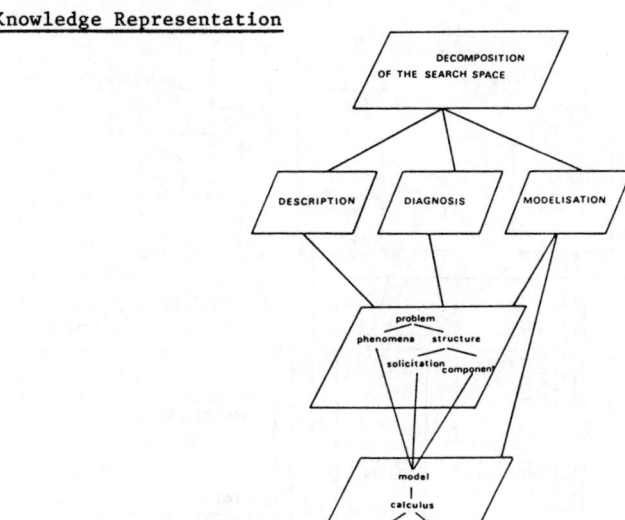

Fig. 1 : Achitecture of the knowledge base

The elementary knowledge can be written in production rules. Two kinds may be separated : Meta-rules and Rules. Meta-rules are knowledge about knowledge and produce the various steps of the strategy which may be identified as : description, diagnosis and modelisation. In this way it is possible to run the developpment of the search space. This space may be represented as a three whose nodes are got from the execution of series of transitions. A potential tree corresponding to all possible enchainments exists. The correct management of this search space means to develop only one part of the tree which contains the solution and which must be as small as possible (Laurent, 1984). Rules contain knowledge specific to each elementary action. Description is displaid in order to give initial data to the system for the different situations which may occure. Diagnosis tasks consists in choosing among a list of values, that's the case of the identification stage and of the selection of a mechanical theory. Modelisation tasks are similar to the diagnosis ones but they also have to generate concepts.

To deal with uncertainty - corresponding to rules of the thumb - a certainty factor is introduced. Objects are also manipulated by the rules. Objects are concepts composed of a property list. Each property may take series of values. Objects are sorted out.

ARCHITECTURE OF THE SYSTEM

It is composed of three main modules (Fig. 2) :

- knowledge base
- inference engine
- data base

and front-end modules

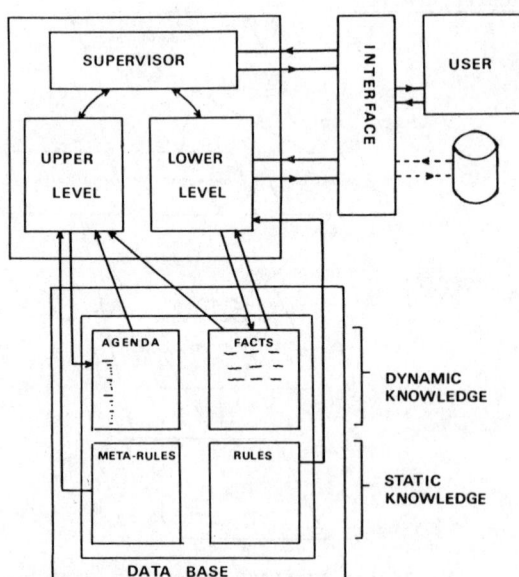

Fig. 2 : Architecture of the system

The knowledge Base

The knowledge base contains two parts : rules and definition of contexts. Rules are represented as on Fig.3.

```
( TYPE   Name
    IF    Premise1  AND
          Premise2  AND
               OR   Premise3
                    Premise4
                ⋮
    THEN   Conclusion1
                ⋮
    ELSE   Conclusionj
                ⋮
                                )
```

Fig. 3 : Sketch of a production rule

Three kinds of rules are recognized with two for meta-rules (decomposition of the search space, partial constraints for decomposition) and one kind for lower level rules. The last ones are organized in set corresponding to elementary action. In each set, backward chaining or forward chaining is notified. Premises are composed of a predicate and of a triple (Parameter Context Value) as in Emycin (Van Melle, 1977). Predicates behave as filter which compares premise triple to facts base triple. About twenty predicates are defined. Two main classes could be distinguished :

- Those which use directly the context normally associated to the premise.
- Those which manipulate contexts in relation with the context associated to the rule.

The first one contains control predicate and comparaison predicate. Control predicates test the state of the facts or of the decomposition strategy advancement.

Example : (KNOWN FAILURE-MODE)

(ACTIVABLE-STAGE   GLOBAL-MODEL)

Comparaison predicates enhance similarity between facts and premise.

Example (GREATER DIAMETER 80)

The second one are divided into :

- Those which associated context relation is bound to the inference engine control.

Example (THOSE-WHO ELEMENT-TYPE OPENING)

- Those which relation is included in the predicate

Example (BOUND ?X   ELEMENT-TYPE   OPENING)

In the last two types, apart from the certainty factor, a series of contexts which satisfy the premises are returned as a free variable. This variable is local to the rule and is often used to conclude.
Conclusions are of two forms, those which add facts to the data base, those which trigger actions to communicate with the user. In the first form, there are those which conclude on a parameter value and those which generate one or more context. The certainty factor is involved only in the case of rules

which conclude on a value. In the second form there are information actions and ask-for information actions.
As said before a context is a set of parameters or properties.

                    Example   SOLLICITATION CONTEXT

                              - type
                              - direction

As concepts are sorted out, liaisons and creation modes (interactive, by rule...) must be specified in their definition. For the parameters, list of possible values, type, how to find them out when a backward chaining is used, informations message are given in their definition.

The Inference Engine

The mechanism which has to operate the knowledge base is called the inference engine. It produces new facts and new actions from an initial data. Two levels are separated (Fig. 2). The first one run meta-rules and produce a sequence of actions (upper level) and the second one executes these actions (lower level). A supervisor manage the connection between the two levels. Depth first search is always used.

The upper level : two forward chaining are successively trigered. The first one uses facts to generate a list of actions in order to progress toward the solution. This list contains no order between actions. The second one uses facts and potential actions to generate partials orders in the list of actions. With the list of actions and the set of partial orders, a total order is created and is added to the agenda. This agenda evolves all along the consultation. No contexts - interactively or by rule - are generated, and nothing is asked to the user at this level.

The lower level : This level works either by forward chaining or by backward chaining. New facts are deduced from existing or asked for ones and contexts generated by rules or interactively with the user.

The data base

The data base contains factual knowledge and deductive knowledge. It is a collection of data structures used by programs in a disciplined way. The modules of the program interact through out the data-base (Charniak, 1980). In production systems, a great part of the total run time is spent in the comparison of patterns to a collection of facts. That's the pattern matching. Execution speed is bound to the implementation of this pattern matching (Forgy, 1982). So, it is located in the central memory and in order to reduce the number of pattern, not all ruleset are present at the same time.

Front End Modules

There are two front-end modules the interpreteur and the interface with the user
The interpreteur : traduces the knowledge base written by the knowledge engineer in something understable by the system. It also has to check the validity of the syntax.

The interface : In order to describe the action to run, to start treatment task and to operate on results, a communication language has been defined. This language acts on algorithm (commands) and on data. With the command a tutorial mode (interactive) or a batch mode can be selected. Actions on data authorize to consult base facts, agenda, rules used ...

EXPERT KNOWLEDGE

The used knowledge is presented through the following example : a pressure vessel anchored to the ground (Fig. 4)

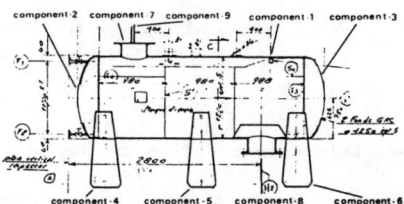

Fig. 4 : The pressure vessel

Figure 5 displays relations between contexts of different nature (as a structure and a component) and between contexts of the same nature (as two components). Two sub-tree may be identified : one for representing the problem and one for representing its modelisation.

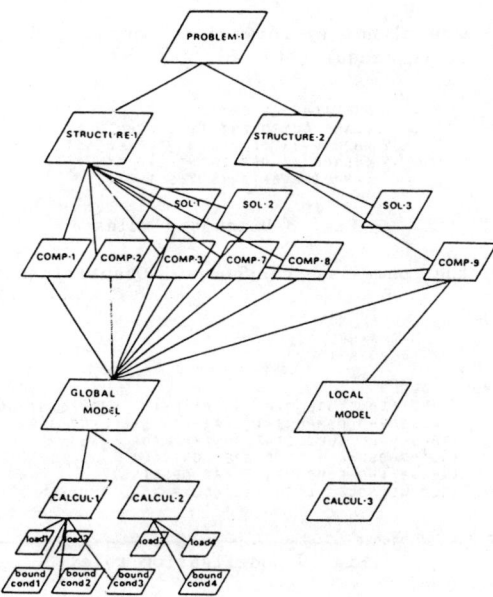

Fig. 5 Relations between context

The decomposition of the search space is shown on Fig. 6

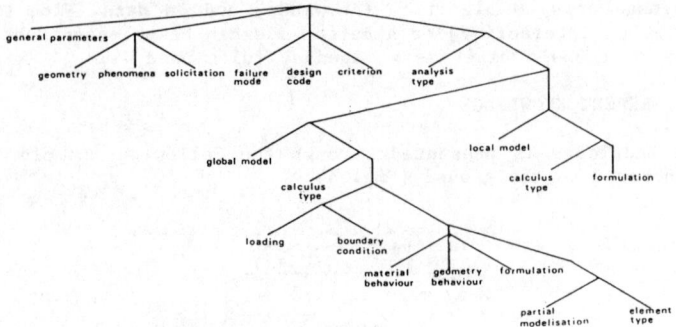

Fig.6 A possible decomposition of the search space

Forward chaining is used for description actions. Rules ask for information or deduce them (fig.7)

```
(:RULE SHELL1                        (:RULE WEIGTH1
    (SAME COMPONENT_TYPE SHELL)          (KNOWN STRUCTURE_TYPE)
==> (ASK-FOR LENGTH)                 ==> (GEN-PNODE (SOLICITATION SOLICITATION_TYPE WEIGTH))
    (ASK-FOR DIAMETER)                   (CHARACTERIZATION STATIC 1.0)
    (ASK-FOR THICKNESS))                 (CONCERNED_FCN NORMAL 1.0))
```

Fig.7 Description rules

Diagnosis can be made either by forward or backward chaining. Rules can use free variable or can be causal (Fig. 8)

```
(:RULE PIPING-EFFORT
    (SAME COMPONENT_TYPE OPENING)
    (BOUND ?X COMPONENT_TYPE CONSTITUENT)
    (VSAME ?X RESULTANT-EFFORTS YES)
==> (APPLIED_EFFORTS YES 1.0))
```

Fig. 8 Diagnosis rules

Modelisation is the same as diagnosis. Quantificators are often used (Fig.9).

```
(:RULESET GLOBAL_MODEL
    BACKWARD
    TEMPORARY

(:RULE MODEL1
    (SAME STRUCTURE_TYPE SHEET_METAL_WORKED_STRUCTURE)
    (THOSE-WHO COMPONENT_TYPE (OR SHELL STIFFENERS SUPPORT))
    (THOSE-WHO SOLICITATION_TYPE (OR PRESSURE WEIGTH))
    (NOT-THOSE-WHO 3D_SHAPE CONICAL)
    (THOSE-WHO COMPONENT_TYPE OPENING)
==> (GLOBAL_MODEL (GEN-PROCEDURE-NODE 'MODEL TOP-NODE) 1.0))
```

Fig. 9 Modelisation rule

The document 1 to 2 illustrate a consultation.

CONCLUSION

The generation of a strategy of analysis is an abstract task. Until now all the decisions behaved to the engineer. Our prototype displays the possibility of such a tool based on artificial intelligence technics. Meanwhile, the following difficulties could be mentionned :

- Those bound to the problem description. For discrete structures like sheet-metal-worked structures, components are perfectly defined. These components contain prototypical notions (A shell : 1 privilegied dimension, bound to heads...) When, massive structures has to be studied, these notions vanish. So it's difficult to describe the object topology, the geometry, the semantic attributes.
- Those to find a compromise among competence and run time : the flexibility of Expert System is paid in return by important CPU time and the needs of a large memory space. So, when response time of the system exceeded widely the user's reflexion time, the consultation become boring. That's particulary sensible when search space is very large - that's our case - if some precautions are not taken. In order to avoid this problem, the search space is split into ruleset. But this suppose that the problem can be partitionned linearly and even in the best case the competence of the system is diminished.
- Difficulties concerning acquisition and validation of knowledge.

Explanation capabilities in order to securise the user and to help the knowledge engineer in the knowledge base validation are under developpment. The knowledge base has to be extended. Further developpment could be the introduction of backtracking possibilities and cost fonctions to replace certainty factor in some heuristic selection rules.

REFERENCES

Buchanan B.G (1982). In Machine Intelligence, Intelligent Systems : pratices and perspectives, Vol 10, J. Wisley Ed.

Charniak E. , C.K. Riesbeck, D.V. Mac Dermott (1980). Artifical intelligence programming , Lawrence Erlbaum, New Jersey.

Forgy C.L. (1982) . Rete : A fast algorithm for the many pattern/many object pattern match problem, J. of Art. Int , 19 , pp. 17-37.

Hervé M. (1983). Technologies avancées en mécanique : calcul de structures, CETIM Information, 80.

Laurent J.P (1984). La structure de contrôle dans les systèmes experts, Technique et science informatique, 3 pp. 161 - 176.

Meziane A. (1986). Système intégré pour la réalisation d'ossatures et charpentes par ordinateur, Thèse 3ème cycle, Université de Valenciennes, France.

Moreau J.P. J. VUILLEMIN (1985). Le projet SYCOMORE, Bulletin de liaison de l'INRIA, 102, pp. 2-10

Rehak D.R, H.C, Harvard, D. Sriram, (1985). In knowledge engineering in C.A.D., Environment for structural engineering applications, North Holland, Amsterdam, pp. 89-118

Reyner M., (1984). Système expert en conception mécanique, 4èmes journées internationales sur les S.E et leurs applications, Avignon.

Van Melle W (1977). A domain-independant system that aids in constructing knowledge-base consultation programs, Report N° STAN - CS-80-20 Stanford.

```
LOCATION OF THE DESIGN ?
        - STRUCTURE
        - COMPONENT
>>STRUCTURE
ADVANCEMENT STAGE OF DESIGN ?
        - BUILT
        - DESIGNED
        - PRELIMINARY_DESIGNED
>>DESIGNED
DOMAIN OF UTILIZATION FOR PROBLEM-1 -- NUCLEAR ETC -- ?
>>NUCLEAR
ENVIRONMENT TYPE ?
        - COMPRESSIBLE_FLUID
        - UNCOMPRESSIBLE_FLUID
        - GROUND
>>GROUND
TYPE OF STRUCTURE-1 ? -- PROBLEM-1 --
        - SHEET_METAL_WORKED_STRUCTURE
        - PIPING_STRUCTURE
        - WELDED_STRUCTURE
        - MECHANISM
        - END
>>SHEET_METAL_WORKED_STRUCTURE
IS STRUCTURE-1 THE MAIN'S ONE ?
>>YES
ARE DISPLACEMENTS LIMITED FOR STRUCTURE-1 ?
>>
TYPE OF STRUCTURE-2 ? -- PROBLEM-1 --
        - SHEET_METAL_WORKED_STRUCTURE
        - PIPING_STRUCTURE
        - WELDED_STRUCTURE
        - MECHANISM
        - END
>>PIPING_STRUCTURE
IS STRUCTURE-2 THE MAIN'S ONE ?
>>NO
ARE DISPLACEMENTS LIMITED FOR STRUCTURE-2 ?
>>
TYPE OF STRUCTURE-3 ? -- PROBLEM-1 --
        - SHEET_METAL_WORKED_STRUCTURE
        - PIPING_STRUCTURE
        - WELDED_STRUCTURE
        - MECHANISM
        - END
>>
FAILURE-MODE TYPE TO BE STUDIED ?
        - EXCESSIVE DEFORMATION
        - PROGRESSIVE DEFORMATION
        - DUCTILE RUPTURE
        - BRITTLE FRACTURE
        - CRACK INITIATION
        - BUCKLING
        - CREEPING
        - RESONANCE
>>
WHICH DESIGN-CODE MUST BE USED -- ASME III ASME VIII CODAP ETC -- ?
    .
    .
    .
```

Doc 1. Illustrative consultation

```
MATERIAL ?
>>METAL
CALCULUS TEMPERATURE ?
>>25
BUCKLING TEMPERATURE -- C -- ?
>>430
TYPE OF SHEET_METAL_WORKED_STRUCTURE FOR STRUCTURE-1 ?
       - TANK
       - PRESSURE_VESSEL
>>PRESSURE_VESSEL
TYPE OF CONTENT FOR STRUCTURE-1 ?
       - GAZ
       - LIQUID
       - LIQUID_WITH_GAZ
>>LIQUID
TYPE OF COMPONENT-1 ? -- STRUCTURE-1 --
       - SHELL
       - HEAD
       - OPENING
       - FLANGE
       - STIFFENERS
       - SUPPORTS
       - OTHER
       - END
>>SHELL
LENGTH OF COMPONENT-1 -- IN MM -- ?
>>2800
DIAMETER OF COMPONENT-1 -- IN MM -- ?
>>1250
THICKNESS OF COMPONENT-1 -- IN MM -- ?
>>10
COMPONENTS BOUND WITH COMPONENT-1 ?
>>COMPONENT-7 , COMPONENT-8

-----------------------------------------------------------
!                                                         !
!                 ANALYSIS STRATEGY                       !
!                                                         !
-----------------------------------------------------------

-->STRUCTURE-1
FAILURE MODE  : GROWTH_DISTORSION (CERTAINTY 0.100E+01)
DESIGN CODE   : ASME8DIV2A4 (CERTAINTY 0.100E+01)
DESIGN CODE   : ASME8DIV2A5 (CERTAINTY 0.100E+01)
CRITERION     : 2RE (CERTAINTY 0.100E+01)
CRITERION     : 1RE (CERTAINTY 0.100E+01)


-->CALCULUS-1
GLOBAL MODEL COMPONENTS : COMPONENT-1 COMPONENT-2 COMPONENT-3 COMPONENT-4
COMPONENT-5 COMPONENT-6  COMPONENT-7 COMPONENT-8 SOLLICITATION-1
SOLLICITATION-2 (CERTAINTY 0.100E+01)
TYPE             : STATIC (CERTAINTY 0.100E+01)
MATERIAL LAW : LINEAR_ELASTIC (CERTAINTY 0.600E+00)
GEOMETRICAL BEHAVIOUR : STABLE (CERTAINTY 0.100E+01)
MODEL            : FINITE_ELEMENTS (CERTAINTY 0.100E+01)
PARTIAL MODELISATION : TRANSVERSAL_PLAN (CERTAINTY 0.600E+00)
- BOUNDARY_CONDITION-1
TYPE             : SIMPLY_SUPPORTED (CERTAINTY 0.100E+01)
FROM             : COMPONENT-4 (CERTAINTY 0.100E+01)
- BOUNDARY_CONDITION-2
TYPE             : SIMPLY_SUPPORTED (CERTAINTY 0.100E+01)
FROM             : COMPONENT-5 (CERTAINTY 0.100E+01)
- BOUNDARY_CONDITION-3
TYPE             : SIMPLY_SUPPORTED (CERTAINTY 0.100E+01)
FROM             : COMPONENT-6 (CERTAINTY 0.100E+01)
- LOADING-1
TYPE             : PRESSURE (CERTAINTY 0.100E+01)
SPACE REPRESENTATION : SURFACE (CERTAINTY 0.100E+01)
- LOADING-2
TYPE             : WEIGHT (CERTAINTY 0.100E+01)
SPACE REPRESENTATION : VOLUME (CERTAINTY 0.100E+01)


-->CALCULUS-2
LOCAL MODEL COMPONENT : COMPONENT-8 (CERTAINTY 0.100E+01)
MODEL            : BIJLAARD (CERTAINTY 0.100E+01)
```

Doc 2. Illustrative consultation

SELECTION AND EVALUATION OF STRUCTURAL OPTIMIZATION STRATEGIES BY MEANS OF EXPERT SYSTEMS

Dietrich Hartmann

University of Dortmund, FRG

ABSTRACT

One of the main obstacles in structural optimization is the appropriate selection of an optimization procedure. Although numerous candidates of strategies are available the selection is very difficult because many algorithms do not solve each problem with the same degree of efficiency as desired. Therefore, an inadequate selection may lead to prohibitive computational effort, in the worst case, no solutions may be obtained. Substantial experience is required to make the adequate decision. In particular, those engineers who are not familiarized with structural optimization have to face major problems. Consequently, a computer oriented system, which is consulting non-experts, would be of extraordinary value. In recent years, so called expert systems or knowledge based systems have been developed which give non-experts the opportunity to use expertise from areas normally requiring an expert with years of training and experience. Such an expert system can be applied to the task to select appropriate optimization algorithms and to streamline an evaluation. Within the scope of this paper it is demonstrated how an expert system can be used to create a knowledge base which permits a computer assisted selection and evaluation of optimization strategies.

THEORETICAL BACKGROUND

The structural optimization problem can generally be described as follows:

determine a vector

$$\mathbf{x} = \left\{ x_1, x_2, \cdots, x_n \right\}^T$$

that minimizes the optimization criterion $Q(\mathbf{x})$
subject to the set of constraints

$$h_j(\mathbf{x}) \geq 0, \; j = 1, 2, \cdots, m$$

where specified constraints are associated with structural analysis equations (e. g. linear finite element stiffness equation, eigenvalue equations, etc).

In the context of structural optimization, the variables in the vector $\mathbf{x}$ may be structural or geometrical design variables. The optimization criterion may be some quantity (e. g. cost or a mechanically oriented entity) which should be optimized. Certain constraints may be quantities which impose limitations on the structural response (stresses, displacements, buckling loads, etc.), other prescribe bounds due to fabrication, design and analysis validity considerations.

In recent time, the evaluation of the structural response is most often based upon a finite element technique. Since the involved matrices (e. g. stiffness matrix, mass matrix, load vector, etc.) are directly dependend on the optimization vector the analysis process has to be multitudinously repeated (re-analysis). In particular, both the necessity of re-analysis and the high nonlinearity of the optimization problem require a sophisticated selection of the appropriate solution technique for optimization. That fact holds even more if a major number of optimization variables has been specified and major structural systems are to be considered.

FIELD OF APPLIFICATION

Basically, two main categories of optimization methods can be distinguished:

- transformation methods
- direct methods.

Within the first category the constrained optimization problem is transformed into a sequence of unconstrained problems by virtue of penalty, barrier or augmented Lagrangian functions. After transformation a wide variety of methods called "hill-climbing-methods" can be applied to the solution of the individual unconstrained problem within the sequence. Hereby, "hill-climbing-methods" are the following:

- search techniques
- gradient or 1st order methods
- Newton or 2nd order methods.

Contrary to the transformation approach direct methods are solving constrained problems. Two subcategories can be devided: either the original problem is solved or a sequence of special constrained subproblems (e. g. linear or quadratic subproblems). The original problem can be solved by means of modified hill-climbing-methods where modifications are necessary because of the active constraints which hinder the hill- climbing process. In recent years, the sequential solution of special constrained subproblems has been proved as particularily effective (e.g. method of reduced generalized gradient through recursive linearization, method of sequential quadratic programming by successive quadratic approximation).

To give a rough survey on the potential methods and techniques the following diagram (see Fig.1) schematically illustrates the coherences.

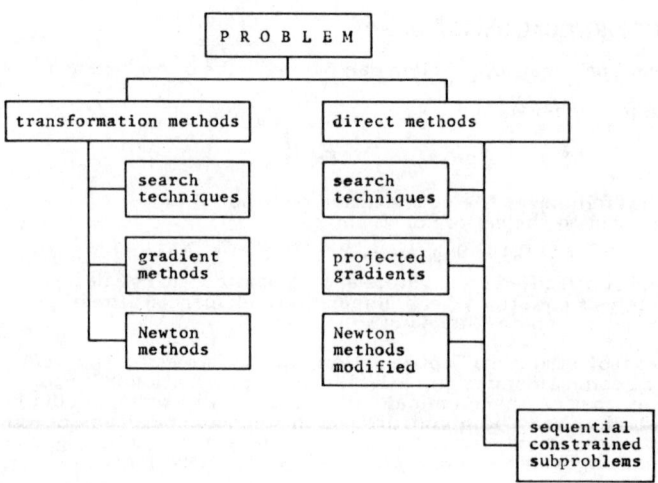

Fig. 1: survey on main categories of methods

The selection and evaluation has to be based upon expertise in the field of structural optimization. That is to say that the essential performance criteria as well as the computational consequences regarding the structural analysis have to be considered. The following performance criteria may be taken into account:

- efficiency in terms of computational effort and storage demand
- general applicability in nonlinear cases
- reliability in practical applications
- local and global convergence behaviour
- sensitivity to changes in the structural optimization model
- expandability with respect to new developments
- stability in degenerated or ill-conditioned cases.

In recent years the computational effort has been viewed as the prime quantity of interest. From the viewpoint of structural optimization, however, further features than solely computation are of significance.

It becomes apparent that it is not an easy task to evaluate all perfomance criteria mentioned above because a diversity of knowledge about the domain of structural optimization is needed. Furthermore, the knowledge has many forms. To give some examples, knowledge may be typified by the following items:

- descriptive definitions of objects and terms as well as relationships between each other
- description of typical situations
- certain and uncertain facts
- rules for decisions subject to constraints
- hypotheses
- conditions
- conclusions

A keystone to the success of an expert system is the effective representation of the domain knowledge into a coherent knowledge base. Besides appropriate representation mechanismi, the expert system must provide versatile reasoning facilities as well as components by means of which a dialogue between the knowledge base and the user can be accomplished.

FUNDAMENTALS OF OBJECT ORIENTED PROGRAMMING

The computer assisted selection is based upon so called production rule languages which provide the complete environment for the design, creation, maintenance and use of a knowledge domain. Similarly to the use of conventional programming languages (e. g. FORTRAN), in the field of knowledge engineering characteristic languages have been developed, like LISP, SMALL TALK, PROLOG, and INSIGHT: After comprehensive evaluation it could be stated that INSIGHT constitutes an excellent language for practical engineering applications. Therefore, the INSIGHT language [1] has been taken in order to design, create and run knowledge bases.

The INSIGHT knowledge system consists of a knowledge base compiler and an inference system. The inference system (INSIGHT) executes the knowledge base created and designed by the knowledge engineer. The knowledge base compiler (PRGEN) compiles the source knowledge base into a compiled version which is run by the inference system. The compiler translates and streamlines the source material so that the run version requires less memory and runs faster than the non-compiled one. During the execution of the run version (consultation phase) the user is automatically questioned for information needed to make conclusions and inferences about the subject, in conformity with the factual and value information as well as facts and goals enbedded into the knowledge base.

The production rule language (PRL) is a format for knowledge description; it gives the prime framework upon which knowledge and information is organized. PRL has a syntax by means of which the logical flow of questions and conclusions can be presented to the user. As one gains additional information and new knowledge, new cognitive items can be added to the knowledge base. PRL makes use of reserved key words (defined by capital letters) and operators to program the necessary statements of the reasoning process. The reserved words are

AND	DISPLAY	IF	RULE
ARE	ELSE	IS	THEN
CF	END	OFF	THRESHOLD
CONFIDENCE	EXPAND	ON	TITLE

These words constitute the fundamental means to write the complete source knowledge base which, at least, has to consist of the following components:

- TITLE-statement to define a title
- THRESHOLD-statement to specify the lowest percentage of confidence
- goal statement to designate all major conclusions and hypotheses
- production rule statements to express the expertise available
- END-statement to define the logical and physical end of the knowledge base.

Within the scope of this paper it is sufficient to discuss only goal and production rule statements.

<u>Goal Statements</u>

A goal statement consists of a phrase or sentence which describes a conclusion that can be reached. At least one goal is mandatory. Each goal must be preceded by a goal outline number which, therefore, may serve to divide the goals into appropriate goal groups. Of course, the goals have to be related to specified production rules in a proper manner, otherwise the knowledge base would not be useful. The following example shows a set of goals and subgoals for a knowledge base which may be used as a means for selecting optimization strategies:

1.	Solution technique is transformation method
1.1	Multiplier Method
1.2	Penalty Function Method
1.3	Barrier Function Method
2.	Solution technique is direct method
2.1	Constrained Search Technique
2.2	Sequential Constrained Technique
2.2.1	Generalized Reduced Gradient
2.2.2	Sequential Quadratic Programming

Either it is possible to identify a narrow area of interest or to evaluate the total goal list. To prove the entire set of goals, or specified ones, goals must correspond with some knowledge about the goals. This knowledge is embedded into rules which are to be discussed in the following subchapter.

<u>Production Rules</u>

Production rules define the expertise about a knowledge domain. Thus, they can be thought of as a way in which descriptive definitions and hypotheses are proved through an inference process, similarly to the philosophy human beings solve problems. As a minimum, production rules have the following general form:

RULE	rule name
IF	supporting condition
THEN	conclusion

Of course, more complicated constructs may be needed for real world situations, e.g. one may specify more than one supporting condition by using the reserved key word AND. Also, multiple conclusions are allowed. Due to the logic flow of a given problem rules can be nested and related to other rules. If supporting conditions cannot be verified another set of conclusions can be made by using the ELSE key word. In addition, facts can be associated with confidence evaluation through THRESHOLD and CONFIDENCE ON/OFF statements, respectively. Both commands are used to quantify the degree of conclusion accuracy. Confidence levels provide a simple means and a way to specify so called "fuzzy knowledge" of real wold situations.

Production rules are capable of representing information of the following type:

- factual information in terms of phrases and sentences
- numerical information in terms of arithmetic assignments and simple conditional statements
- set information by means of which individual rules can be processed as a coherent set
  (key word "IS" and "ARE" define set type information)

In particular, the last mechanism allows to increase the efficiency and to make the knowledge base more sophisticated because group of facts can be designed which significantly streamlines the inference process. The following examples demonstrates the set type information:

```
RULE    one point pattern line search
IF      derivatives available ARE 2nd order
THEN    use one point pattern line search

RULE    three point pattern line search
IF      derivatives available ARE none
THEN    use three point pattern line search
```

Since the key word ARE is used both rules are grouped together. The decision which type of techniques (one point or three point pattern line search) is recommended only depends on evaluating the information following the word ARE.

Finally, it should be pointed out that the key word EXPAND permits to provide additional information displayed on the screen only when requested. The DISPLAY command provides a means to automatically display text. Both types of commands may be necessary to improve the understandibility of the reasoning process.

Consequently, the elements outlined above provide the necessary tools for engineers to describe the relationships between facts, events, details and qualities that comprise the knowledge domain with respect to structural optimization. Contrary to conventional (or procedural) programming where the engineer must describe precisely "how a result is to be computed" in logic programming it is described "what it is that must be solved". In other words: using the INSIGHT language, instead of asking "What is the algorithm that will solve my problem?" the engineer asks "What are the facts and rules which describe my problem?". As a consequence, the INSIGHT language provides a natural way to specify knowledge about a domain of interest in a logical format for inferences.

PROGRAM DESCRIPTION

According to the requirements in conventional programming, a modular approach is adopted to code the knowledge base. Within the scope of this contribution only two individual goals are being considered (as space is limited). Each of both goals is associated with a collection of specified rules defining cause and effect relationships within the knowledge domain of the corresponding goals. The two goals persued are the following :

- a rapid problem classification of the problem in view
- a rapid preliminary selection of an appropriate optimization method in view of the structural optimization problem to be solved.

Within each rule collection pathes consisting of one or more rules have to be constructed by the knowledge engineer according to the goal persued. Pathes within the knowledge base may be interpreted as conditions that only arise when such rules are to be validated that refer further rules. With regard to conventional programming, the collection of rules necessary to persue specified goals can be thought of as "subroutines" in logic programming. Therefore, similar aspects as in modular programming of conventional modules (e.g. subroutines, functions) apply to the coding of "rule pathes" within the knowledge base. Of course, rule pathes may communicate with each other like program units do in "classical programming". It is here that the interconnection between goals and rules become significant.

In order to illustrate the use of rule pathes, an elementary knowledge base is presented in the appendix which is based upon several distinct rules pathes for each goal. However, interconnections of the two main goals are not considered here.

Rapid Problem Classification

The first goal "rapid problem classification" is associated with four rule pathes (path 1a, 1b, 1c, 1d, 1e; path 2a, 2b, 2c, 2d, 2e; etc.). In each of the four pathes it is checked

- which type of optimization criterion
- which type of constraints

can be specified for the problem in view. When evaluating a specified rule path the expert system questions for information according to the logic structure of the rule collection. Due to the information provided, interferences and conclusions with respect to the category of the optimization problem (e.g. unconstrained algorithmically nonlinear optimization problem) are made. In addition, a recommendation is given on which type of solution techniques is to be taken (e.g. derivative free method). The dependencies between the four rule pathes along with the potential conclusions are schematically demonstrated in the figure 2.

Rapid Preliminary Selection

The second set of rule pathes solves the problem to rapidly select the best optimization strategy among four alternatives. The four competitive methods are the following :

1. The evolution stategy (EVOL) as a repesentative
   method for direct solution. The solution process is based
   upon a stochastic search technique associated with
   versatile in-built adaptation mechanismi [2], [3], [4].

2. The method of multipiers (MOM) as a representative
   method for solution through the transformation approach.
   The method repesents a combination between the
   Lagrangian approach and the penalty function method [5] [6].

3. The method of generalized reduced gradients (GRG) as a
   representative method for the direct methodology.
   This method handles the optimization problem directly
   through sequential linearization of the original
   optimization criterion and/or the constraints [6], [7].
   constraints [6], [7].

4. The method of sequential quadratic programming (SQP)
   which, again, is a repesentative method for the direct
   approach. This method approximates the original problem
   through a sequence of constrained quadratic
   subproblems [6], [7], [8].

All methods considered currently provide very advanced and sophisticated techniques within the area of structural optimization.

The selection is based upon two groups of performance criterion where the first one is considered to be of higher rank than the second. The evaluation refers to expertise and knowledge about the numerical behaviour of the four competitors, gathered in recent years. Also, comperative studies and tests like that one performed by SCHITTKOWSKI [8] has been taken into account. For each of the stategies in view, a rule path has been created (Rules EVOL 1, 2.1, 2.2, 3.1, 3.2; Rules MOM1, 2.1, 2.2, 3.1, 3.2; etc) which identifies the prime qualities of the corresponding optimization strategy.

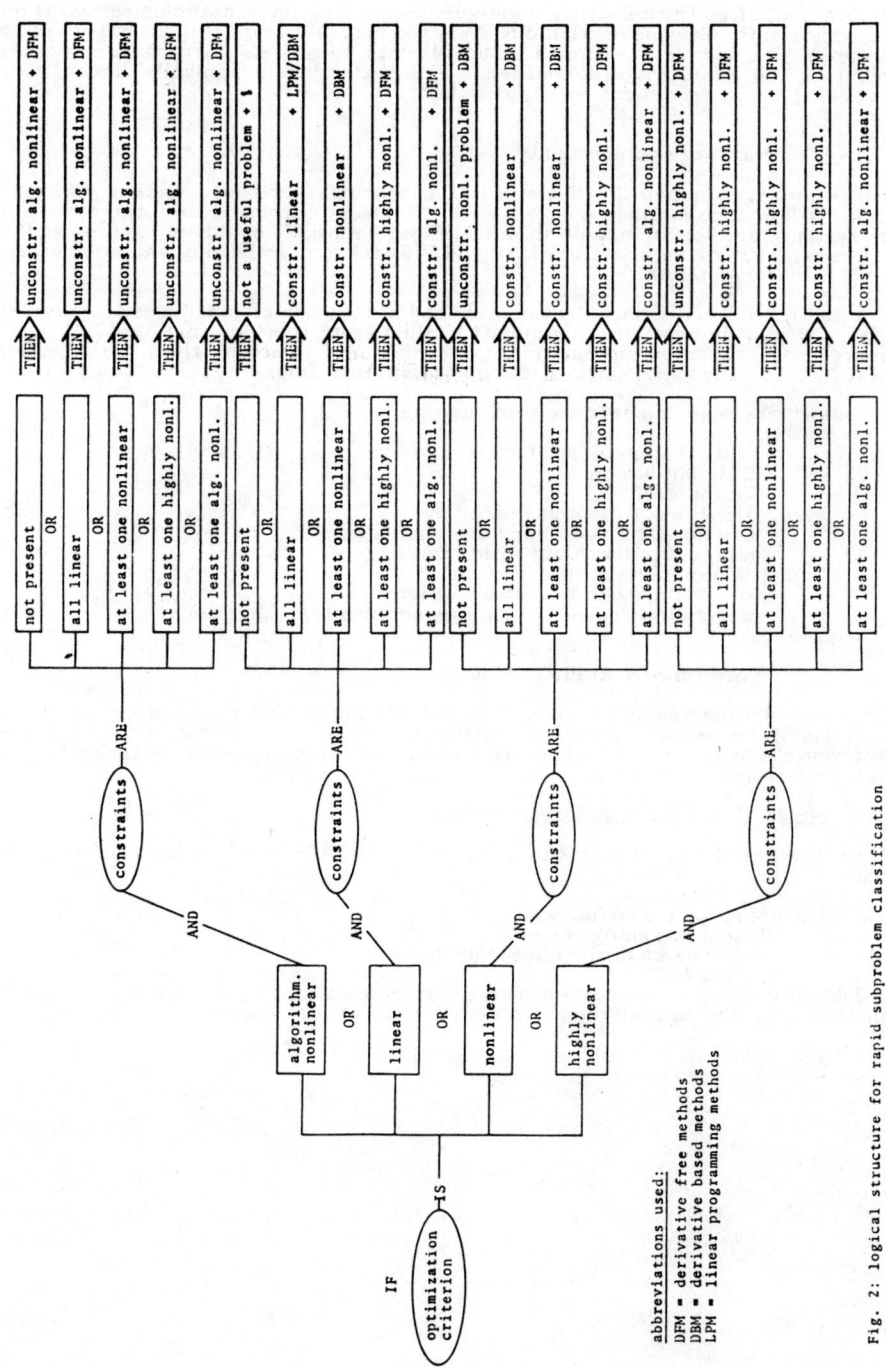

Fig. 2: logical structure for rapid subproblem classification

In addition, after the selection is completed, a check upon pathological cases and numerical difficulties is carried out. As is the case in the classification part of the program the user is questioned in an attempt to get sufficient information for reaching the appropriate conclusions. The next figure graphically demonstrates the rule path interconnection and structure.

Hardware compatibility

The knowledge base outlined above is a microcomputer-based system. Since the INSIGHT language is taken as a basis for knowledge representation the drawbacks of systems developed in well known LISP and PROLOG could be overcome: Most often expert systems written in LISP and PROLOG tend to be slow and cumbersome on the microcomputer.

Due to the INSIGHT philosophy the overhead of interpreted languages such as PROLOG and LISP is avoided. This quality results from the fact that INSIGHT uses a source code compiler (non-standard, mini-PASCAL) to shrink the knowledge-base text files into their most disk- and memory-efficient form.

The hardware requirements are the following:

- IBM PC and compatibles or
  DEC Rainbow or
  Victor 9000
- Hard-disk (recommended) or
  two double-sided floppy disk drives or
  a high density floppy disk drive
- minimum RAM : 448 KByte
- operating system : DOS 2.0 or later
- peripherals required : display and printer, other none.

EXAMPLES OF APPLICATION

The two following examples are to demonstrate the necessary actions a user has to perform if he wishes either to classify a problem considered or to select an appropriate solution technique. The complete knowledge-base-program can be gathered from the appendix.

Example #1: Rapid Problem Classifications

First, the knowledge base has to be loaded (from DOS level) by the following command:
    C > insight
The program responds as follows:
    load a knowledge base
    enter name of the knowledge base
    : paris & CR
(**Bold** information represents messages or prompts sent by the system; underlined material represents user replies or descriptions of user actions.)

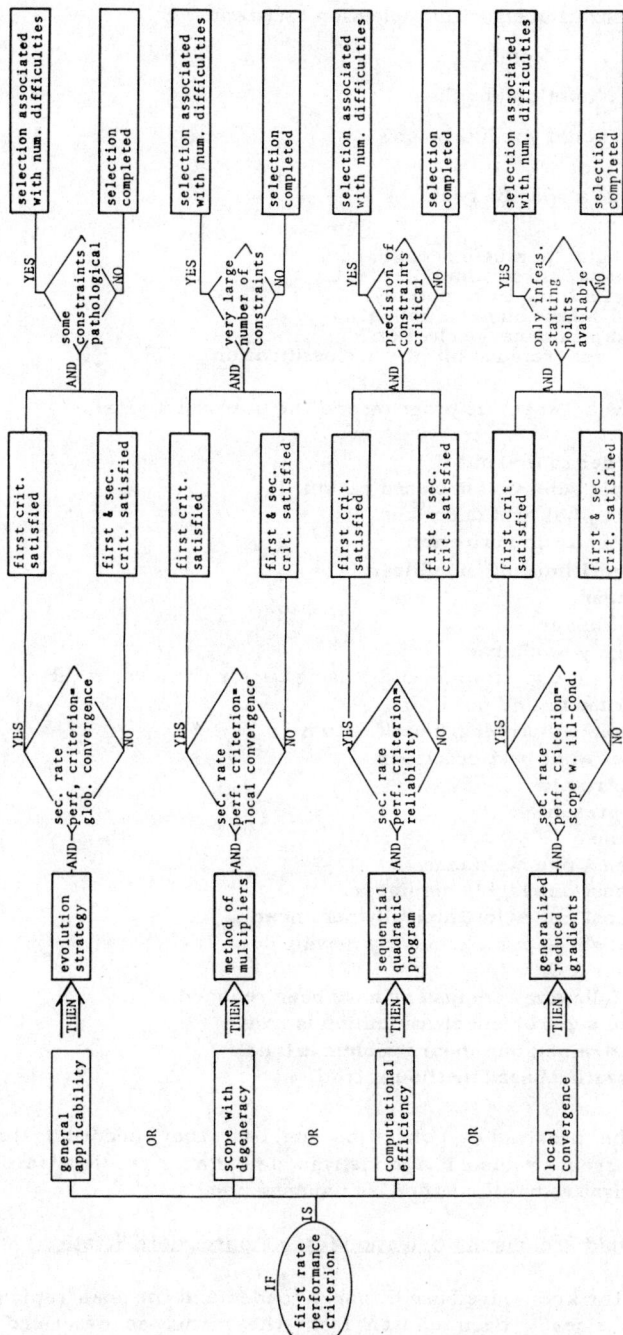

Fig. 3: logical structure for rapid preliminary selection

After loading, a specific goal can be persued by the following process:

      optimization algorithm selection assistant

        .... 5 Menu ....

press function key F5

what would you like to do?
......
......
persue a specific goal;
......
......
highlight "persue a specific goal"
by using "cursor down key" & CR
select a topic to be persued:
rapid subproblem classification
rapid preliminary selection
highlight "rapid subproblem classification" & CR

Now the dialogue between the program and the user is activated;

    in order to find out if:
      rapid subproblem classification
    select what best describes:
      optimization criterion
      algorithmically nonlinear
      linear
      nonlinear
      highly nonlinear
    highlight e.g. "nonlinear" by using cursor down key & CR
    in order to find out if:
      rapid subproblem classification
    select what best describes:
    constraints
    not present
    all linear
    at least one nonlinear
    at least one highly nonlinear
    at least one algorithmically nonlinear
    highlight e.g. "all linear" by moving down the
    cursor & CR
    the following conclusions have been reached
    rapid subproblem classification is true.
    Constrained nonlinear problem is true.
    Derivative based method is true.

As a result, the program is consulting the user that, according to the given answers, the current problem is a "constrained nonlinear problem" for the solution of which a "derivative based method" is recommended.

<u>Example #2: Rapid Preliminary Selection (of an Appropriate Strategy)</u>

Assuming that the knowledge base has been loaded and the goal "rapid preliminary selection" has already been chosen from the menu, as described above the system/user - dialogue starts with the following question:

>   in order to find out if:
>   evolution strategy is one of 4 candidates
>   select what best describes:
>   first rate performance criterion:
>   general applicability
>   scope with degeneracy
>   computation efficiency
>   local convergence.

If the user is uncertain which one of the four performance criteria may best fit his current problem he is enabled to request additional information through a simple EXPAND key stroke (F4). Let us assume that the user desires to favor the quality " general applicability". To give an example, after highlighting the selected property, the EXPAND key stroke provides the following message:

>   more information about
>   general applicability
>
>   **general applicability**
>   designates the ability to solve a wide variety of real
>   world problems despite the degree of nonlinearity
>   incorporated into the numerical model

Choosing the "GO BACK" key allows continuation of the selection process. Provided that the user keeps to the feature "general applicability" he hits CR - key. Hence, the next information or query, respectively, appears onto the screen:

>   ################################
>   # evolution strategy has been selected.#
>   ################################
>   in order to find out if:
>   selection completed
>   is it true that
>   second rate criterion is global convergence?
>   true
>   false

As previously outlined, the user may request additional information about the query "is it true that second rate criterion is global convergence?" before he replies to the question. In the current example it may be assumed that the answer is "true". Highlighting "true" and CR gives the next step within the rule path:

>   in order to find out if:
>   rapid preliminary selection
>   is it true:
>   some constraints may represent pathological situation?
>   true
>   false

If the user again does not know what is meant by the term "pathological" additional information is obtainable. In this example, the decision is made such that no pathological constraints have to be apprehended. Thus, "false" is choosen. As a result, we have the following conclusion:

> **the following conclusions have been reached**
> **evolution strategy is one of 4 candidates is true**
> **first criterion satisfied is true**
> **selection completed is true**
> **first & second criterion satisfied is true**
> **rapid preliminary selection is true**

Consequently, the evolution strategy as a very significant method within the category of search methods (derivatives free methods) has been varified as an appropriate solution technique.

CONCLUSION

The program described above represents a reservoir of knowledge about an indicated domain of structural optimization. It may be considered as an exemplification of a more sophisticated real-world expert system. Obviously, the program system is a backward-chaining or goal driven inference system because it starts from a specified goal and works backward, looking for evidence that supports or contradicts the goals considered.

Without doubt it can be stated that expert systems and knowledge engineering are going to open new directions in the field of engineering applications. Obviously, it is only a matter of time when further domains like finite-element applications, CAD-problems, complicated modelling of structural optimization systems, etc. will be incorporated into corresponding "intelligent" expert systems.

REFERENCES

[1] INSIGHT (1985). Insight Knowledge System Manual. Level Five Research. Inc. . Melbourne Beach, Florida, U.S.A.
[2] Hartmann, D. (1984). Computer Aided Numerical And Structural Optimization By Means Of Evolution Strategies, Report No. UCB/SESM - 84/7. Department of Civil Engineering, University of California, Berkeley.
[3] Schwefel, H. P. (1981). Numerical Optimization of Computer Models. John Wiley & Sons, Ltd., Chichester - New York - Brisbane - Toronto.
[4] Hartmann, D. (1984). Structural Optimization Of Discrete Systems Represented By Finite Elements, Report No. UCB/SESEM - 84/8. Department of Civil Engineering, University of California, Berkeley.
[5] Pierre, D. A. and Lowe, M. J. (1975). Mathematical Programming via Augmented Lagrangians. Addison - Wesley Publishing Company, London - Amsterdam - Sydney - Tokyo.
[6] Morris, A. J. (1982). Foundations of Structural Optimization - A Unified Approch. John Wiley & Sons, Chichester - New York - Bribane - Toronto - Singapore.
[7] Reklaitis, G. V., Ravindran, A. and Ragsdell, K. M. (1983). Engineering Optimization. John Wiley & Sons, Chichester - New York - Bribane - Toronto - Singapore.
[8] Schittkowski, K. (1980). Nonlinear Programming Codes. Lecture Notes in Economics and Mathematical Systems. Springer Verlag, Berlin - Heidelberg - New York.

COMPUTER ASSISTED SELECTION AND EVALUATION
OF STRUCTURAL OPTIMIZATION
ALGORITHMS

TITLE Optimization Algorithm Selection Assistant DISPLAY

Optimization algorithm selection assistant

by

Dietrich Hartmann

University of Dortmund
West Germany

This selection assistant for the selection of numerical
optimization routines has been created to help engineers
not familiarized with the selection of appropriate solu-
techniques for structural optimization to make adequate
decisions.

The knowledge base generated represents the currently
known facts, rules and principles of four significant
algorithms which compete with each other. Since only elementary knowledge is applied the following knowledge base
has to be considered as a pre-version of a more detailed
one.

THRESHOLD = 30

CONFIDENCE OFF

1. rapid subproblem classification
2. rapid preliminary selection

Rules for rapid problem classification

Rule path for algorithmically nonlinear optimization criterion

RULE 1a
IF optimization criterion IS algorithmically nonlinear
AND constraints ARE not present
THEN rapid subproblem classification
AND unconstrained algorithmically nonlinear problem
AND derivative free methods

RULE 1b
IF optimization criterion IS algorithmically nonlinear
AND constraints ARE all linear
THEN rapid subproblem classification
AND unconstrained algorithmically nonlinear problem
AND derivative free methods

RULE 1c
IF optimization criterion IS algorithmically nonlinear
AND constraints ARE at least one nonlinear
THEN rapid subproblem classification
AND unconstrained algorithmically nonlinear problem

AND derivative free methods

RULE 1d
IF optimization criterion IS algorithmically nonlinear
AND constraints ARE at least one highly nonlinear
THEN rapid subproblem classification
AND unconstrained algorithmically nonlinear problem
AND derivative free methods

RULE 1e
IF optimization criterion IS algorithmically nonlinear
AND constraints ARE at least one algorithmically nonlinear
THEN rapid subproblem classification
AND unconstrained algorithmically nonlinear problem
AND derivative free methods

Rule path for linear optimization criterion

RULE 2a
IF optimization criterion IS linear
AND constraints ARE not present
THEN rapid subproblem classification
AND not a useful problem

RULE 2b
IF optimization criterion IS linear
AND constraints ARE all linear
THEN rapid subproblem classification
AND constrained linear problem
AND linear programming methods
AND eventually, derivative based methods

RULE 2c
IF optimization criterion IS linear
AND constraints ARE at least one nonlinear
THEN rapid subproblem classification
AND constrained nonlinear problem
AND derivative based methods

RULE 2d
IF optimization criterion IS linear
AND constraints ARE at least one highly nonlinear
THEN rapid subproblem classification
AND constrained highly nonlinear problem
AND derivative free methods

RULE 2e
IF optimization criterion IS linear
AND constraints ARE at least one algorithmically nonlinear
THEN rapid subproblem classification
AND constrained algorithmically nonlinear
AND derivative free methods

Rule path for nonlinear optimization criterion

RULE 3a
IF optimization criterion IS nonlinear
AND constraints ARE not present
THEN rapid subproblem classification

AND unconstrained nonlinear problem
AND derivative based methods

RULE 3b
IF optimization criterion IS nonlinear
AND constraints ARE all linear
THEN rapid subproblem classification
AND constrained nonlinear problem
AND derivative based methods

RULE 3c
IF optimization criterion IS nonlinear
AND constraints ARE at least one nonlinear
THEN rapid subproblem classification
AND constrained nonlinear problem
AND derivative based methods

RULE 3d
IF optimization criterion IS nonlinear
AND constraints ARE at least one highly nonlinear
THEN rapid subproblem classification
AND constrained highly nonlinear problem
AND derivative free methods

RULE 3e
IF optimization criterion IS nonlinear
AND constraints ARE at least one algorithmically nonlinear
THEN rapid subproblem classification
AND constrained algorithm. nonlinear problem
AND derivative free methods

  Rule path for highly nonlinear optimization criterion

RULE 4a
IF optimization criterion IS highly nonlinear
AND constraints ARE not present
THEN rapid subproblem classification
AND unconstrained highly nonlinear problem
AND derivative free methods

RULE 4b
IF optimization criterion IS highly nonlinear
AND constraints ARE all linear
THEN rapid subproblem classification
AND constrained highly nonlinear problem
AND derivative free methods

RULE 4c
IF optimization criterion IS highly nonlinear
AND constraints ARE at least one nonlinear
THEN rapid subproblem classification
AND constrained highly nonlinear problem
AND derivative free methods

RULE 4d
IF optimization criterion IS highly nonlinear
AND constraints ARE at least one highly nonlinear
THEN rapid subproblem classification
AND constrained highly nonlinear problem
AND derivative free methods

RULE 4e
IF optimization criterion IS highly nonlinear
AND constraints ARE at least one algorithmically nonlinear
THEN rapid subproblem classification
AND constrained algorithm. nonlinear problem
AND derivative free methods

  Rules for rapid preliminary selection

  Rule path for evolution strategies (EVOL)

RULE EVOL1
IF first rate performance criterion IS general applicability
THEN evolution strategy is one of 4 candidates
AND DISPLAY EVOL1 conclusion
AND first criterion satisfied

RULE EVOL2.1
IF evolution strategy is one of 4 candidates
AND second rate criterion is global convergence
THEN selection completed
AND first & second criterion satisfied

RULE EVOL2.2
IF evolution strategy is one of 4 candidates
AND first criterion satisfied
THEN selection completed

RULE EVOL3.1
IF evolution strategy is one of 4 candidates
AND selection completed
AND some constraints may represent pathological situation
THEN rapid preliminary selection
AND numerical difficulties may occur
AND WARNING!! evolution strategy may prematurely terminate
AND WARNING!! optimum may not be obtained

RULE EVOL3.2
IF evolution strategy is one of 4 candidates
AND selection completed
THEN rapid preliminary selection

  Rule path for method of multipliers (MOM)

RULE MOM1
IF first rate performance criterion IS scope with degeneracy
THEN method of multipliers
AND DISPLAY MOM1 conclusion
AND first criterion satisfied

RULE MOM2.1
IF method of multipliers
AND second rate criterion is local convergence
THEN selection completed
AND first & second criterion satisfied

RULE MOM2.2
IF method of multipliers
AND first criterion satisfied

53

THEN selection completed

RULE MOM3.1
IF method of multipliers
AND selection completed
AND a very large number of constraints occurs
THEN rapid preliminary selection
AND numerical difficulties may occur
AND WARNING!! slow convergence may occur
AND WARNING!! optimum may not be obtained

RULE MOM3.2
IF method of multipliers
AND selection completed
THEN rapid preliminary selection

!---!
!   Rule path for sequential quadratic programming (SQP) !
!---!

RULE SQP1
IF first rate performance criterion IS computat. efficiency
THEN sequential quadratic programming
AND DISPLAY SQP1 conclusion
AND first criterion satisfied

RULE SQP2.1
IF sequential quadratic programming
AND second rate criterion is reliability
THEN selection completed
AND first & second criterion satisfied

RULE SQP2.2
IF sequential quadratic programming
AND first criterion satisfied
THEN selection completed

RULE SQP3.1
IF sequential quadratic programming
AND selection completed
AND precision of constraints is difficult to control
THEN rapid preliminary selection
AND numerical difficulties may occur
AND WARNING!! numerical problems may occur
AND WARNING!! optimum may not be obtained

RULE SQP.2
IF sequential quadratic programming
AND selection completed
THEN rapid preliminary selection

!---!
!   Rule path for generalized reduced gradients (GRG) !
!---!

RULE GRG1
IF first rate performance criterion IS local convergence
THEN generalized reduced gradients
AND DISPLAY GRG1 conclusion
AND first criterion satisfied

RULE GRG2.1
IF generalized reduced gradients
AND second rate criterion is scope with ill-conditioning
THEN selection completed
AND first & second criterion satisfied

RULE GRG2.2
IF generalized reduced gradients
AND first criterion satisfied
THEN selection completed

RULE GRG3.1
IF generalized reduced gradients
AND selection completed
AND only infeasible starting points are available
THEN rapid preliminary selection
AND numerical difficulties may occur
AND WARNING!! only feasible initial solutions converge
AND ADVISE!! to overcome that problem, take evol. strategy

RULE GRG3.2
IF generalized reduced gradients
AND selection completed
THEN rapid preliminary selection

!---!
     information automatically printed onto the screen
!---!

DISPLAY EVOL1 conclusion
****
*   evolution strategy has been selected   *
****

DISPLAY MOM1 conclusion
****
* method of multipliers has been selected   *
****

DISPLAY SQP1 conclusion
****
* sequential quadratic programming has been selected *
****

DISPLAY GRG1 conclusion
****
* generalized reduced gradients has been selected *
****

!---!
           additional information on request
!---!

EXPAND algorithmically nonlinear
    "algorithmically nonlinear"
    indicates that the optimization criterion or the con-
    straint considered, respectively, cannot be represented

in terms of a conventional function but only in an algorithmical format.

example:
algorithm - compare the four values
            A,B,C,D nonlinearly depending
            on the optimization variables;
            then:
            take the absolutely greatest one
            as that value which is to be
            minimized;

!---------------
EXPAND linear

"linear"
indicates that the optimization criterion or the constraint considered, respectively, is represented in terms of a function in which the optimization variables occur in a linear form.

example: $f(x_1,x_2) = x_1 + x_2$

!---------------
EXPAND nonlinear

"nonlinear"
indicates that the optimization criterion or the constraint considered, respectively, is represented in terms of a function in which the optimization variables occur in a nonlinear form. However, in contrast to the specification "highly nonlinear" the term "nonlinear" has the property that 1st and 2nd order derivatives of the corresponding functions can be established with ease.

example: $f(x_1,x_2) = x_1^2 + x_2^2$
            $df/dx_1 = 2 \cdot x_1$
            $df/dx_2 = 2 \cdot x_2$

!---------------
EXPAND highly nonlinear

"highly nonlinear"
indicates that the optimization criterion or the constraint considered, respectively, is represented in terms of a function in which the optimization variables occur in a highly nonlinear form. However, in contrast to the specification "nonlinear" the term "highly nonlinear" denotes that 1st and 2nd order derivatives of the corresponding functions can not be established with ease.

example: $f(x_1,x_2) = 1/\sqrt{(\sin x_1)} + \ldots$
            $df/dx_1 = \ldots$
            $df/dx_2 = \ldots$

!---------------
EXPAND derivative free methods

derivative free methods (DFM)
are methods which are based upon search directions and search step calculations without information on 1st and 2nd derivatives of neither the optimization criterion nor the constraints. Derivative freee methods are also called search methods.

!---------------
EXPAND derivative beased methods

derivative based methods (DBM)
are methods which make use of 1st and 2nd order derivatives of either the optimization criterion or the constraints, respectively. If the derivatives have to be calculated by means of numerical difference formulas the computational effort may be significantly increased. Representative methods are the following:

gradient type methods
Newton or Newton like methods

!---------------
EXPAND linear programming methods

linear programming methods (LPM)
are methods which, according to the linear form of the optimization criterion and the constraints, are capable to solve huge problems, with thousands of variables and constraints. Representative methods are the "simplex method", modern programs use the closely related "revised simplex method". Both methods utilize matrix and vector theory and notation.

!---------------
EXPAND general applicability

general applicability
designates the ability to solve a wide variety of real world problems despite the degree of nonlinearity incorporated into the model.

!---------------
EXPAND scope with degeneracy

scope with degeneracy
represents the capability to solve problems which are typified by means of redundant active constraints at the optimal solution. That is to say that, in the worst case, the constrained solution is identical with the unconstrained solution of the optimization criterion. In the case of Lagrangian functions, using Langrange multipliers, degeneracy can be realized through zero Lagrange multipliers. The same effect occurs if a wide range of magnitude is covered by the Lagrange multipliers.

Degeneracy can be controlled if an automatic selection of active constraints along with the rejection of nonactive is provided. In particular, if a major number of constraints are inactive the dynamic selection of the active set of constraints significantly may increase the speed of the solution process. Such situations often appear within the optimization of structural systems.

!---------------

EXPAND computational efficiency

    computational efficiency
is the "classical quality" most often considered as the key criterion for numerical optimization. The term itself involves two further sub-qualities which are convergence speed and storage requirements. Both are of main significance for numerical considerations. However, it is emphasized that in structural optimization additional aspects have ton be taken into account, like reliability, global convergence , etc..

EXPAND local convergence

    local convergence
typifies the ability to rapidly find the nearest local optimum with respect to the starting point of the solution process. That property is of importance if , through test runs, an excellent approximation of the optimum is already available.

EXPAND second rate criterion is global convergence

    global convergence
is the ability to convergence towards the best optimum among at least two local optima. That quality is very important in structural optimization because most often numerous local optima may occur in practical design problems.

EXPAND second rate criterion is local convergence

    local convergence
typifies the ability to rapidly find the nearest local optimum with respect to the starting point of the solution process. That property is of importance if , through test runs, an excellent approximation of the optimum is already available.

EXPAND second rate criterion is reliability

    reliability
represents a crucial property that indicates a high number of problems successfully solved, despite the type of non-linearity of the optimization criterion and constraints, respectively.

EXPAND second rate criterion is scope with ill-conditioning

    ill-conditioning
is very closely related to the condition number of the Hesse matrix of the Lagrangian function used. If the condition number is very small the Hesse matrix becomes indefinite. Such type of problems are difficult to manage. In particular, the ill-conditioned problems may be associated with transformation methods using penalty or barrier functions.

EXPAND some constraints may represent pathological situation

    pathological constraints
are constraints which contours run nearly in parallel to the contours of the optimization criterion. Hence, the region in which more successful points could be expected is narrowed.

EXPAND a very large number of constraints occurs

    large number of constraints
indicates the situation in which the number of constraints in relation to the number of optimization variables exceeds a sparficied value. To give a guiding principle for a large number:

    total number of constraints > 10 · total number of optimization variables

EXPAND precision of constraints is difficult to controll

    constraints precision
is a quality by means of which feasible and infeasible solution are effected. If the accuracy of corresponding constraints is crucial with respect to the acceptance of a solution the method must be capable to controll the defect between the feasible and the infeasible domain.

END

EXPERT SYSTEMS IN MECHANICAL ENGINEERING DEMONSTRATED BY THE SELECTION OF FLEXIBLE COUPLINGS

A. Spielvogel, W. Platt and Ch. Troeder

*Institut für Maschinenelemente und Maschinengestaltung, RWTH Aachen,
Schinkelstrasse 8, D-5100 Aachen, FRG*

ABSTRACT

In this article the background and the actual implementation of expert systems in mechanical engineering is discribed. Details are documented by examples from the selection and dimensioning of flexible couplings.

KEYWORDS

Artificial intelligence, expert systems in mechanical engineering, flexible couplings.

INTRODUCTION

Expert Systems can be used as a powerful design tool in mechanical engineering. The engineer often uses a heuristic approach, when he selects and combines machine parts during the design phase. Due to this fact the qualitiy of the solution depends strongly on the knowledge of the engineer. If he would be able to use a knowledge based system, which stores all relevant data known about a specific problem, an optimum solution would be guaranteed. This technique will be demonstrated by a software package, which supports the selection of flexible couplings.

The selection is not restricted to static load condition but although the dynamic behavior of the whole driving system must be considered. So the design process can become a very complex and time consuming task, because the procedure must be repeated for many different couplings.

This work can better be done by a computer program, which was developed by the institute of machine elements and machine design (RWTH Aachen). It is written in PROLOG, but some parts, especially the graphic output is supported by GKS and FORTRAN. It can be used by a designer without special knowledge about the selection and dimensioning of flexible couplings.

THEORETICAL BACKGROUND - THE EXPERT SYSTEM APPROACH

In mechanical engineering a solution technique as shown in figure 1 is used. For a specific task of problems a catalog of specifications will be defined in which the main goal and the boundary conditions are described. From this catalog a model will be created, which represents the task and its boundary conditions and restrictions in a sufficient manner. This work must be done by an expert and requests a lot of experience. Even smallest errors in modelling are followed by a failure of the problem solution. In the next step, algorithms are introduced to describe the model and to do the simulation. If the behavior is convenient, a solution is received. In the other case the whole procedure or a part of it must be repeated.

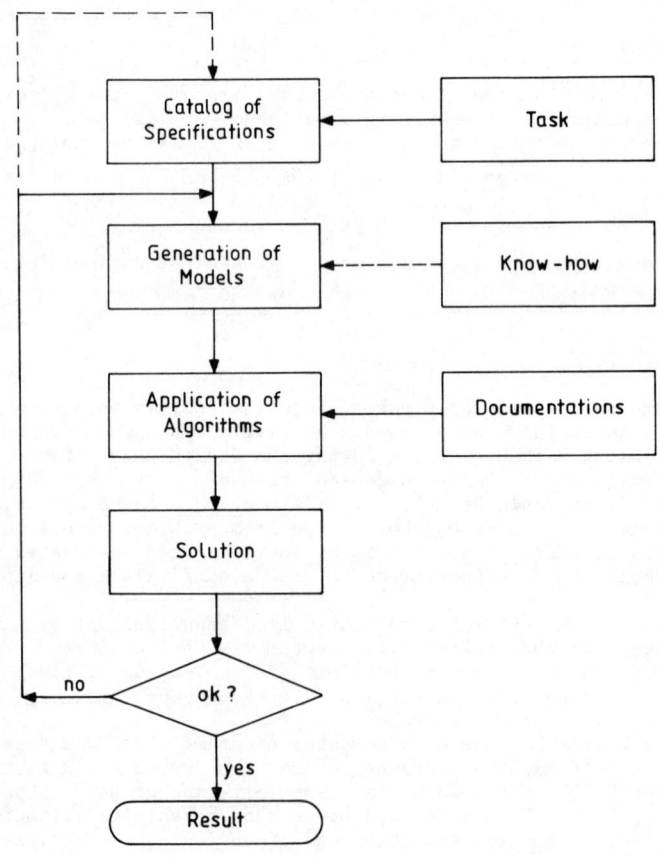

Fig. 1.

With conventional programming technique only the algorithmic part of the solution process can be supported. Tools from artificial intelligence (AI) allows it, to go a step further and to introduce computer systems in modelling and reasoning. One tool from AI is the expert system.

Expert systems are software packages, which store the specialized competency of an expert in the form of knowledge (facts, which can be found in literature) and experience (general rules, analogy, personal criteria etc). With associated rules they are able to generate problem solutions for given data (Com & Pro, 1985).

How the support by an expert system can take place may be demonstrated by figure 2.

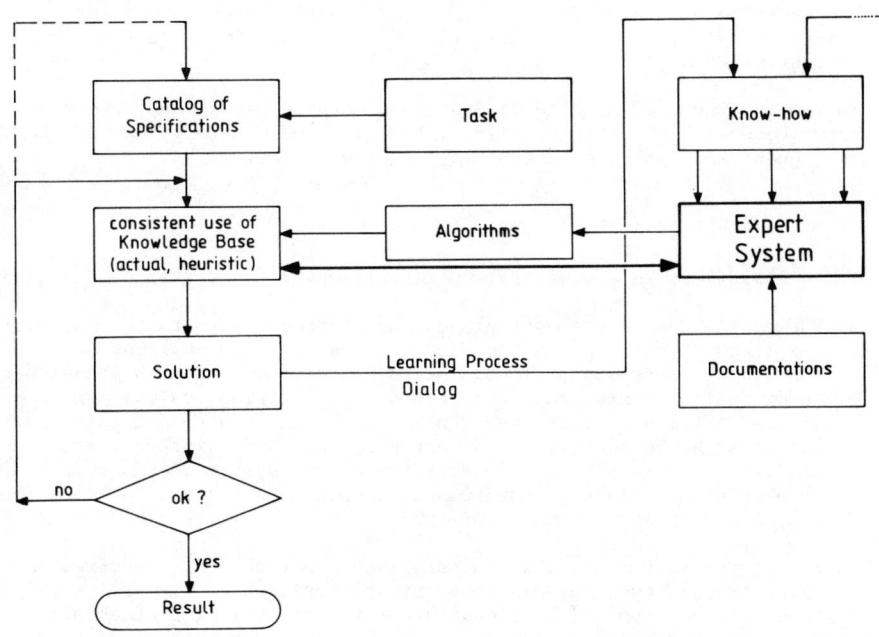

Fig. 2.

The expert system stores the know-how of a human expert and learns from solutions, which have been accepted by the user. It although has access to the documentation as a sort of information. The documentation can be made transparent to the user. By this way the user can follow the decision making of the system and learn from it. The strategy for finding a solution can include an algorithmic part as for example some numerical operations, graphics or simulation.

The learning process of the expert system is named "knowledge acquisition" and allows updating and modifying the current knowledge base.

The Coupling Expert was designed to support the optimum selection of
flexible couplings in rotordynamics. It includes the following criteria:

* static and dynamic behavior of the system
* static reliability of the coupling
* dynamic reliability of the coupling
* security against thermal overloading
* stiffness of the coupling
* misalignment of the shafts
* assembly problems
* costs

The software is written in PROLOG (Schnupp, Schmauch, Leibrand, 1984) and integrates a graphic capability written in FORTRAN and GKS.

The underlaying theory for the algorithmic part (calculations for dimensioning) are done with DIN 740 and some extensions described in (Benner, 1984).

As the Institute of Machine Elements and Machine Design has some years of experience with flexible couplings, the required knowledge on this domain is present and has been integrated into the system.

FIELD OF APPLICATION

There are several areas of application, in which the system can be used:

a) During the design phase of plants the engineer wants to test several couplings in a short time to find the best for his application concerning costs and performance. In practice its to much manual work to dimension several couplings for one application, so the engineer will stop the selection, when he has found one convenient solution. But it might be reasonable to test couplings from different producers or different typ. This is done by the expert system, because it has an open database with some hundreds of couplings different in dimensions, typ, producer and characteristics.

b) For the producer it may be interesting to use the expert system in sales. When a customer has some special application, the salesman, who does not need special knowledge about the dimensioning of couplings, can present several different solutions in a short amount of time to the customer.

c) During education the student can learn how to dimension and select a coupling for a special task. This is supported by the "explanation component", which allows the user to ask for reasons for a question in each phase of the consultation.

In general expert systems can be devided into several categories (Rehak, Fenves, 1984).

* interpretation systems
* prediction systems
* diagnosis systems
* debugging systems
* design systems
* planning systems
* repairing systems
* tutoring systems
* control systems

So there is a wide future range of applications for expert systems in mechanical engineering.

PROGRAM DESCRIPTION

The system is designed as a dialog oriented software package. A "help"-facility exists and the ability to ask the system in each step, which answers are possible. Following feature are build-in:

* enlargement of the database concerning new couplings, drivings and load
* definition of an enviroment
* selection of all couplings, which are sufficient for a load case
* comparison concerning price, optimum layout or assembly
* display of all calculated values and characteristics in combination with the plant
* graphic display of the static and dynamic graph of the coupling, security limits concerning load, misalignment and local wants.

Some typical results from the last point are shown in figure 3 to 5.

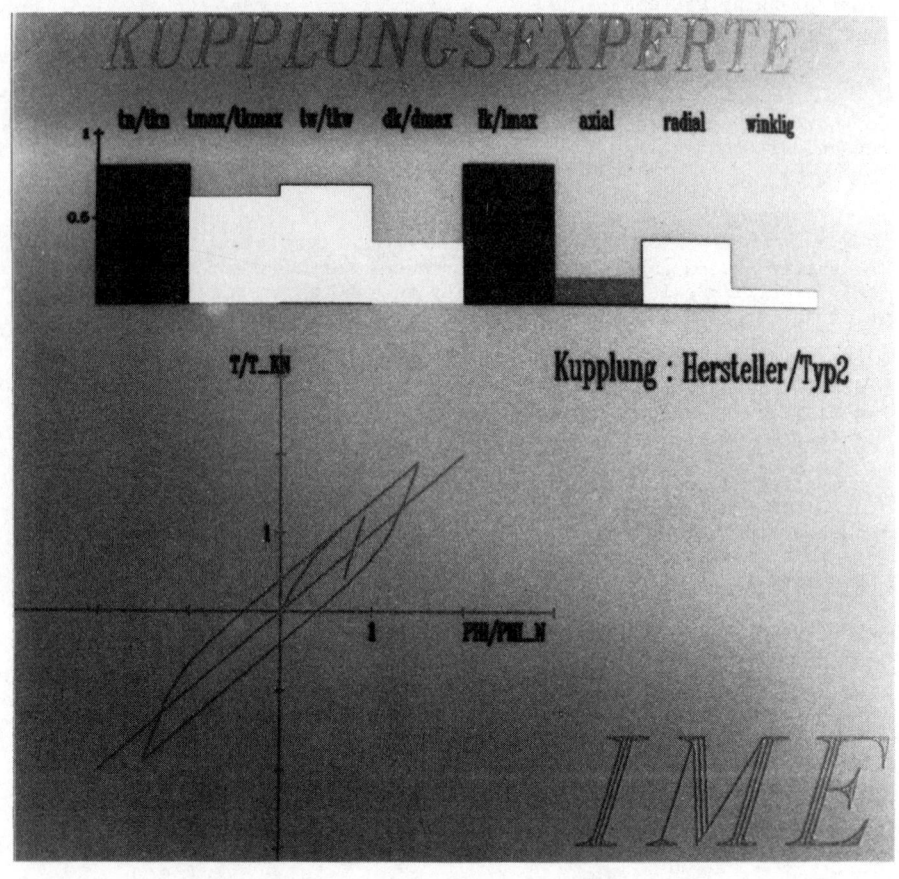

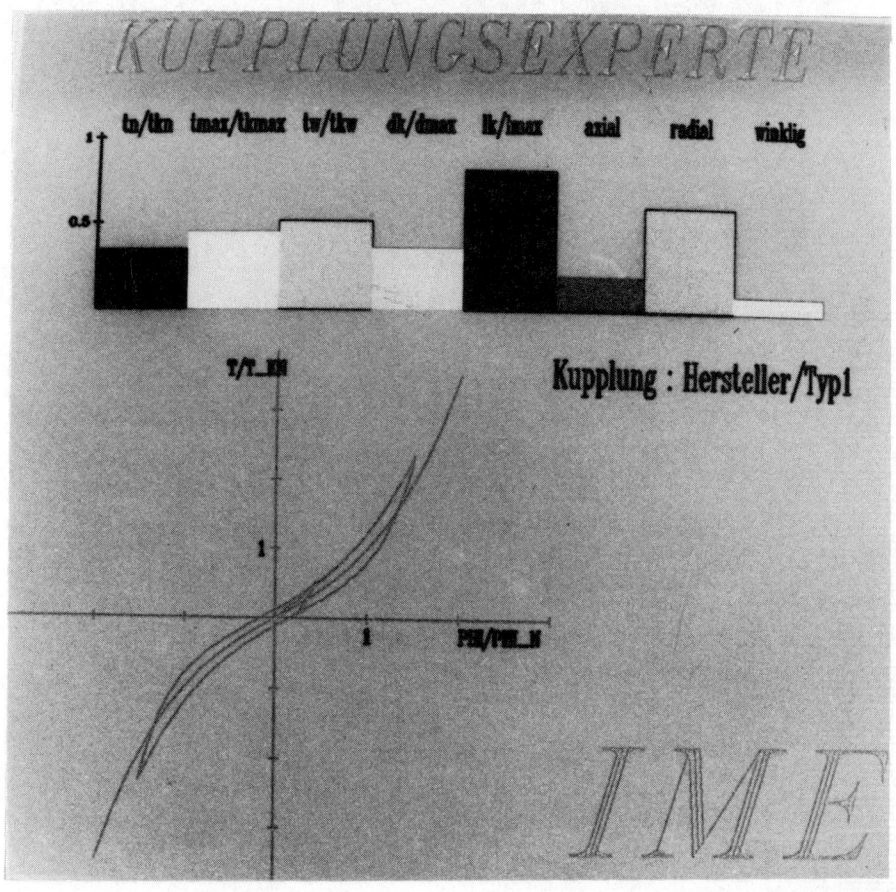

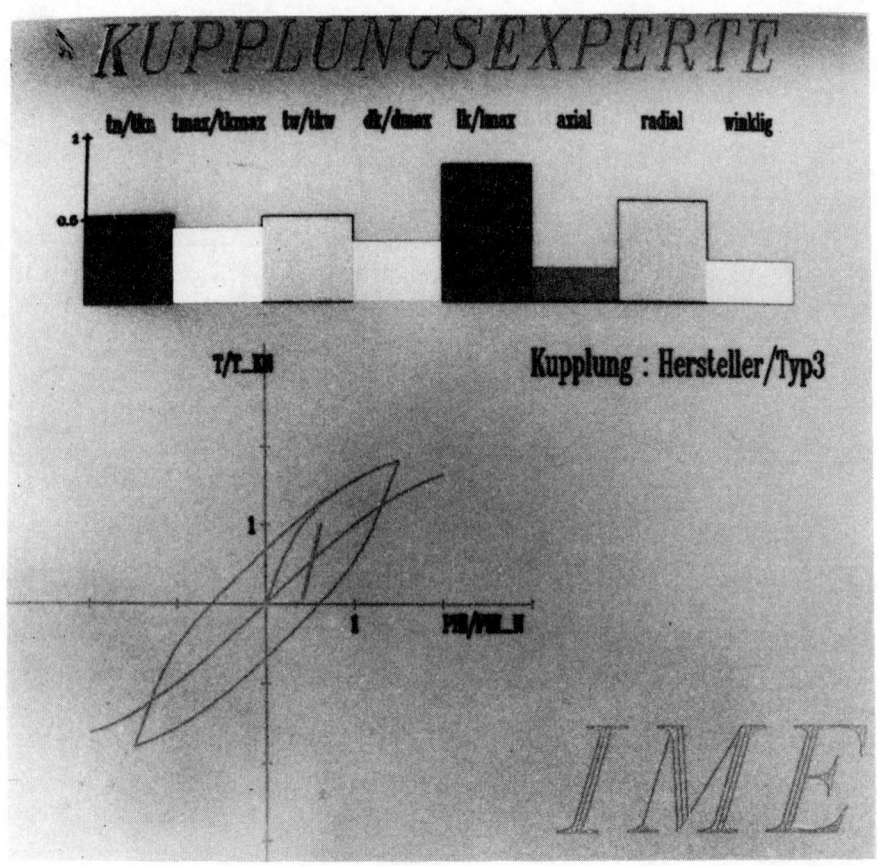

HARDWARE COMPATIBILITIES

At the moment the Coupling Expert is installed on a Data General workstation DS4200 with 4 Mb of memory and 140 Mb of disk storage. The workstation is equiped with a coloured graphic display. The software requirements are a C-Compiler (Prolog is written in C), and FORTRAN with GKS for the graphics. The Prolog Interpreter is supplied by Interface Computer, Munich and can be installed on all common computers, which have an Unix operating system.

EXAMPLE OF APPLICATION

Now an example including input and output is described:

On the drive side (engine)
    a three phase current motor (type 160M4)
    output   11 KW
    speed of rotation   1450 Rev/min
    torque   72,5 Nm
    moment of inertia   0,0735 Kgm^2

On the load side
    ventilator
    torque   68 Nm
    moment of inertia   5 Kgm^2

Enviroment data
    axial assembly
    maximum diameter   500 mm
    maximum length   500 mm
    axial misalignment   1 mm
    radial misalignment   1 mm
    angular misalignment 1°
    number of starts/hour   150 times
    enviroment temperatur $40^\circ C$
    maximum price   1000 DM

The system presents for this case study three different couplings. From these the coupling with the minimum nominal torque is selected. It has the following main characteristics (there are some more, which are not mentioned because of their special relationship to the dimensioning of couplings).

Coupling
    nominal torque   160 Nm
    maximum torque   480 Nm
    alternating torque   53.33 Nm
    diameter   178 mm
    length   130 mm
    axial misalignment   4 mm
    radial misalignment   4 mm
    angular misalignment 8°
    axial and radial assembly
    price   400 DM
    type   stromag pna 16

The coupling has a slight progressive hysteresis and good securitys (see figure 3). There is no danger of resonances, because no eigenfrequency is close to frequency of operation.

LITERATURE

Benner, J. (1984). Mathematisches Modell zur Beschreibung des dynamischen Verhaltens drehnachgiebiger Wellenkupplungen, <u>VDI-Berichte Nr. 524</u>

<u>Com & Pro</u>, Sonderpublikation Elektronik 16, 1.8.85

Rehak, D.R., Fenves, S.J. (1984). Expert Systems in Construction, <u>Computers in Engineering</u> Vol. 1, p. 228-235, ASME

Schnupp, P., Schmauch, G., Leibrandt, U. (1984). Was ist Prolog?, <u>Elektronische Rechenanlagen 4</u>

Chapter 2
FINITE AND BOUNDARY ELEMENT METHOD

AN APPROACH TO THE QUALITY ASSESSMENT OF FINITE ELEMENT MESHES

T. K. Hellen

CEGB, Berkeley Nuclear Laboratories, Berkeley, UK

ABSTRACT

A computerised approach to assessing the quality of finite element mesh data is described. Such data can be assembled and checked for accuracy by a range of pre-processor codes, either interactively or in batch mode. However, the shape of individual elements in such meshes can have a considerable effect on results, and yet high degrees of distortion cannot be picked up by data checking alone. The program module BERQUAL assesses the degree of distortion and other aspects of isoparametric-type elements in two- and three-dimensions and shells, based on the transformation methods used in the theory. The features of this program are described and illustrated with examples which demonstrate how numerical errors can become significant with badly designed meshes.

KEYWORDS

Finite elements; mesh generation; distorted shapes; Jacobian transformations; quality assessment.

INTRODUCTION

The finite element method is used extensively in present day analysis and design, particularly in the power industry where a wide range of stress and thermal analysis is regularly performed on structures, varying from simple two-dimensional tests to large three-dimensional configurations. In addition to the proven success of the finite element method to render accurate results for complicated material behaviour in such components, the practical feasibility of the method has been revolutionised by the advent of interactive graphics, whereby mesh generation and results digestion can be rapidly and, therefore, economically, processed. Although relatively easy to use, such pre- and post-processing programs are internally complicated and not individually of wide general scope, so that in practice a given organisation may use several such programs, each being utilised in applications where its suitability is particularly strong. A common analysis program can be used in conjunction with these.

In the BERSAFE finite element system, several generation programs may be chosen prior to the main analysis step, including generally available external codes as well as its own general pre-processing software. The average user cannot be adequately familiar with every such program, and so meshes that he generates, particularly in three-dimensions, may not always be as good as he thinks they are. For instance, the elements so generated may be of poor aspect ratio, or unduly distorted to fit overall shapes, or with inverted topology. Although comprehensive checking for correctness of data is usually available, a stage is still required to check the mesh and individual elements for such items as shape distortion, skewness, aspect ratio, integration counts for zero energy modes, and adequacy of boundary conditions, all of which cannot be properly covered in the mainstream programs. Such a program, BERQUAL, is described, which is based on properties of the Jacobian determinant calculated throughout each element for isoparametric type families, and is executed on the complete data emanating from any pre-processor that may have been used. A variety of elemental and global checks are made and compared to user-defined tolerances.

JACOBIAN TRANSFORMATIONS

Isoparametric element families are used extensively in finite element applications, covering such arbitrary geometric shapes as two-dimensional, three-dimensional, and plates and shells. The elements share the property that their geometry can be expressed by the same shape functions as the assumed displacement variation, although the functions could differ for the present context. The important property is that a transformation space is used so that an arbitrary shape in the user space $R(x,y,z)$ becomes a unit cube of side lengths of two units in this transformed space $T(\xi,\eta,\zeta)$ (Irons and Ahmad, 1980). The transformation for each point in the element is governed by the mapping function quantified by the Jacobian determinant $\det(J) = \partial(x,y,z)/\partial(\xi,\eta,\zeta)$. In two dimensions, the third co-ordinate is dropped.

Regular shaped elements (e.g. rectangles, parallelopipeds, etc) have a constant value of $\det(J)$ over the element, but as the shape becomes more distorted the order of $\det(J)$, which is a polynomial in (ξ,η,ζ), increases to a limit depending on the order of the shape function polynomials. The element stiffness matrix contains quantities accumulated at the sampling (or Gauss) points and contains a contribution $[\det(J)]^{-1}$ so that if $\det(J)$ is not constant in the element, no exact integrating rule exists. The complete rule which applies to the element when $\det(J)$ is constant is, however, assumed to apply in this case.

If singularities exist in the mapping, $\det(J)$ becomes zero, such as when a side length becomes zero, $\det(J)$ being zero at both vertex nodes and any nodes along that edge. If midside nodes are moved to the quarter points as in crack tip applications, $\det(J)$ is zero at the vertex node identified at the tip. Although in this case enhanced accuracy occurs, in many cases a zero $\det(J)$ would cause significant local errors. Even non-zero variations of $\det(J)$ over the element can give such errors. Hence the BERQUAL program calculates and prints out the value of $\det(J)$ at many points in each element in a manner described subsequently. The ratio of the maximum to minimum value over all such points in the element is a very useful guide to the severity of the mapping distortion, and is therefore one of the key parameters produced.

As a scalar quantity, $\det(J)$ offers much in quality assessment, but does not cover all situations. For instance, very skew elements such as parallelopipeds or elongated rectangles have a constant $\det(J)$ but can give rise to erroneous results. However, a complete assessment of the element shape can be obtained using a vector definition of $\det(J)$.

Using vectors, det(J) at any point in the element equals the vector product of vectors r_ξ, r_η, r_ζ which, for quadrilateral or hexahedral (brick) type elements in two and three dimensions, lie in the directions ξ, η, and ζ as defined in the transformed space T, or along tangents

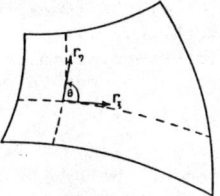

Fig. 1 Definition of tangent vectors and skew angle per reference point

to the lines ξ=const, η=const, ζ=const in the user space R (Fig. 1). Thus,

$$\det(J) = r_\zeta \cdot r_\xi \times r_\eta$$

r_ζ being the out of plane unit vector or the third tangent for two and three dimensions, respectively. The relative magnitude of the vectors r_ξ, r_η, r_ζ defines the aspect ratio at that point, or the <u>tangent vector aspect ratio</u>. The largest divided by the smallest is a representative quantity of interest. The angle, or angles in three dimensions, between these vectors give the local <u>skew angle</u> or angles. The angle used is $90°-\theta$ and in three dimensions the smallest angle of the three is retained as the worst skew angle.

As stated above, det(J) is a polynomial in (ξ, η, ζ), and is easily obtained algebraically for relatively simple shapes. Thus, for the six noded triangle, as long as the sides are straight with midside nodes in the mid position, det(J) is always constant. For four noded elements, or eight-noded again with mid-nodes in the mid positions, we have

$$\det(J) = \frac{1}{16}\left[A + B\xi + C\eta\right]$$

where, as shown in Fig. 2, $A = 4 \times$ area, $B = 4(\Delta_{123} - \Delta_{124})$, $C = 4(\Delta_{134} - \Delta_{124})$, Δ_{ijk} = area of triangle ijk.

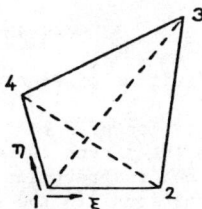

Fig. 2 det(J) for quadrilateral elements

Thus, det(J) is constant if $\Delta_{123} = \Delta_{124}$ and $\Delta_{134} = \Delta_{124}$, i.e. a rectangle or parellelogram and equals $\Delta/4$. It is zero if nodes coalesce, e.g. if 3 and 4 are coincident, $B = 0$ and $C = -A$, so det(J) = 0 when $\eta = 1$, i.e. at nodes 3 and 4. If 3 nodes lie in a line, e.g. 143, $B = 4\Delta_{234}$ and $C = -4\Delta_{124}$, with $A = B = C$. Hence det(J) = 0 if $\xi = -1$, $\eta = 1$, at node 4. Further discussion on the behaviour of det(J) is given by Hellen (1976).

OVERALL ELEMENT ASPECT RATIO AND CURVATURE CHECKS

The variation of det(J) over each element, expressed in the forms above, give a complete spectrum of behaviour for any amount of distortion. However, extra tests are worthwhile to highlight specific features of the elements. In particular, an indicator of the overall chord aspect ratio of the element, as opposed to the detailed point-by-point values derived from the vector form of det(J), and the side distortion for curved quadratic elements, are worthwhile.

For each element, the length of each side is calculated and printed, together with perimeter length, expressed as the sum of all sides in three dimensions as well as two dimensions, each side to perimeter ratio, smallest and largest ratios, and each side's aspect ratio. These ratios are expressed as the straight distance (or chord) between vertex nodes (Fig. 3). In addition to this, it is also necessary to check the curvatures of the side. Thus, if a side contains nodes A, B and C (Fig. 4) the distance DB and angle with respect to DC is printed whenever r = DB is greater than DC by the input parameter DISNOD (default 0.01), where D is the midpoint of AC. Thus, (r, θ)

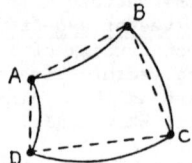

Assume CD > CB > AB > AD

Maximum chord aspect ratio = CD/AD

Fig. 3 Element chord aspect ratio definition

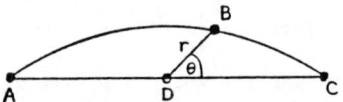

D = midpoint of AC ≡ proper position   BD = (r, θ)

Fig. 4 Definition of side curvature

gives an indication of the distortion of a side. Erroneous local results can accrue as r increases, although the effect is worst for θ ~ 0 or π. If θ = 0 and r = 1/2DC, then the well known quarterpoint node exists which gives rise to displacement variations proportional to the root of the distance from C, as used about elastic crack tips. In other problems, as shown below, this can give rise to large local errors. For $\theta = \frac{\pi}{2}$, symmetric curvature exists and quite large values of r can be used to model boundaries as appropriate, without inducing serious errors. For instance, a curved piece of shell, modelled by a twenty-node quadratic brick element (Fig. 5) can have a value of r, equal at nodes B and B' and up to about DC in length and, with $\theta = \frac{\pi}{2}$, be a very accurate shell element. In this case, aspect ratios of the order of 10 to 1 would apply.

The above discussion covers two- and three-dimensional elements and shells in the program, including quadratic and cubic sided elements. The curvature diagnostic (r,θ) is printed out as indicated, and bears an indirect relation to det(J) as described earlier. The aspect ratio, however, is quite independent and if greater than about 5 to 1 the user should be aware of the usage in his mesh. Thus, in a hypothetical problem where, in a region covered by elements of aspect ratio 10 to 1, if equal stress (or field) gradients exist in those two directions, the model is effectively 10 times more refined in one direction than the other. So high aspect ratios should be introduced when some knowledge of expected stress gradients indicates that this would be efficient.

Ill-conditioning due to high aspect ratios is not usually of significance on modern computers.

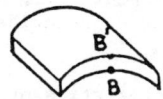

Fig. 5 Example of curved sides

TESTS FOR ZERO ENERGY MODES

In the evaluation of the stiffness matrix of isoparametric and related elements, numerical integration is used, the order of which depends on the order of element being considered. A "complete" integrating rule exists, when, for an undistorted element whose Jacobian determinant is constant throughout the element, the stiffness matrix is evaluated exactly. A "reduced" rule is normally one order less than this. For quadratic displacement elements, a 3x3 or 3x3x3 rule for two and three dimensions, respectively, is complete, whilst 2x2 or 2x2x2 is reduced. The reduced rules have the advantage of often giving very accurate results because of an effect which relieves excessive bending constraints, but occasionally zero energy modes can result. The rule, or order, is specified as an input quantity NGAUS in the BERSAFE system, e.g. NGAUS=2 denotes 2x2 or 2x2x2 as required.

A test for this is to multiply the total number of integration points in the structure by the rank of the stress-strain matrix (i.e. the number of components in the energy vector) to give the "total stiffness rank". This number should exceed the number of free degrees of freedom (the total less the number of precribed degrees of freedom). Failure to do so indicates not enough contributions to the total potential energy have been made and so surplus (or zero) energy modes exist, which can give spurious results. The BERQUAL program works out these values based on the user-specified value of NGAUS. If the rule is complete and the test is passed, it is repeated with the relevant reduced rule for the user's benefit. If the user-specified rule is reduced and the test fails, then it is repeated with the relevant complete rule.

PROGRAM IMPLEMENTATION

Overall Design

The ideas described in the above sections have been implemented in the program step BERQUAL, which is executed after all data, from any data

generation system, has been collected together, but prior to the main
analysis step. It is assumed that the normal data checking associated with
data generation programs has already been satisfactorily conducted. Certain
parameters are defined as input quantities, with suitable defaults, which act
as tolerances for the various characteristics of the elements as described
above.

The overall function of the program is, firstly, to record the contents of
the overall mesh, e.g. number of nodes, numbers and types of boundary
conditions, and perform the overall integration counts to establish if zero
energy modes might exist. Then a loop through each element of topology is
conducted. For each element, the various characteristics are investigated.
Its overall properties are printed, such as material type and element type,
density and weight, then over an array of points controlled by the user, the
det(J) values, tangent vector aspect ratios and skew angles are determined.
The maximum to minimum ratio of det(J) is established over the element,
together with the overall chord aspect ratio and any curved sides, and in all
cases values which exceed the user - prescribed tolerances are highlighted
with suitable warning messages.

A summary list of the main characteristics of each element is printed with,
finally, a summary of maximum and minimum values, ratios and totals, where
applicable.

Main Parameters Investigated
───────────────────────────

The main parameters defined on input as tolerances for the various element
characteristics, are: NINT, ASPECT, RATIOJ, DISNOD, SKEW, NEDGE, NJAC, NSKEW
and NTANG.

The parameter NINT is used to define the number of reference points in each
element, for the evaluation of det(J), the tangent vector aspect ratio, and
the skew angle. The actual number given is for one direction, and the points
are assumed to be equally spaced between the sides in the transformed space
(ξ,η,ζ) (Fig. 6) and are processed in that order. Thus, NINT=9 includes
sufficient points to include mid and quarter point locations. Thirdside
nodes are covered by NINT=7. NINT=9 implies 9x9 in 2D and shells, 9x9x9 in
3D. Owing to the large number of points in 3D, it is advisable to use a
smaller value for NINT, even though less sampling points for det(J) will be
available.

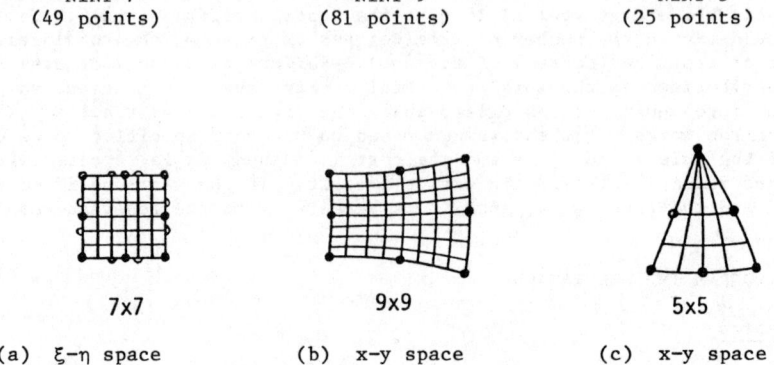

    NINT=7   NINT=9   NINT=5
  (49 points)   (81 points)   (25 points)

     7x7   9x9   5x5
(a) $\xi-\eta$ space   (b) x-y space   (c) x-y space

Fig. 6 Definition of NINT reference points in a 2D element

Depending on the value of NINT used, a large number of reference points may exist per element, particularly in 3D, and some significant CPU time and computer output may appear for large numbers of elements. Hence, input options exist to suppress such effects whilst still enabling suitably large values of NINT to be specified. Firstly, NEDGE allows reference points to be defined only along element sides and faces if 3D, with a further option to suppress within these faces. This is because extreme values of det(J) usually occur on element boundaries. In such cases, the vast majority of the reference points which exist within the element can be avoided. The order of the points is as before, i.e. along ξ, then η, then ζ in the dimensionless space, with points missed out as necessary if interior and NEDGE indicates suppression.

Whatever value of NEDGE is used, output of the reference point quantities det(J), skew angle, and tangent vector aspect ratio can be selectively printed out, at every nth point, using the respective indicators NJAC, NSKEW and NTANG.

The input quantity RATIOJ is the tolerance for maximum to minimum det(J) ratio, which is derived from each reference point covered by the input quantities NINT and NEDGE.

SKEW is the tolerance for the skew angles, for each such reference point. The departure from 90° is considered (i.e. 90° − θ) and in 3D the worst of the three angles at the point is used.

ASPECT pertains to the aspect ratio of each element, defined in two independent ways. Firstly, at each reference point covered by NINT and NEDGE, the tangent vector aspect ratio of Fig. 1 is considered, to give a point-by-point assessment of distortion. Secondly, a global impression of the element is given by comparing the length of each element side and finding the largest divided by the smallest. Straight distances only are considered where the side is curved, so that the quantity is termed the <u>chord aspect ratio</u> of the element (Fig. 3). As a tolerance, ASPECT applies to both definitions of aspect ratio.

DISNOD is the tolerance for distance of midside or thirdside nodes which are disturbed away from their proper positions, expressed as this distance divided by the linear distance of the proper position to the nearest adjacent vertex node. This ratio is r = DB/AD in Fig. 4, the angle θ being defined as <CDB.

DEMONSTRATIVE EXAMPLES

The various element shape checks described can be conducted on any mesh and, if any of the characteristic tolerances are exceeded, suitable warning messages appear. The amounts of distortion can affect the results considerably although no hard and fast rules can be devised. The detailed shapes can depend on local stress gradients, loading, and also on the element type and kind of structure.

The following examples are intended to highlight how errors can accrue with increasing distortion.

<u>Increasing Chord Aspect Ratios</u>

The effect of increasing the elements (chord) aspect ratio, for instance, can in simple examples be shown to induce errors whilst the det(J) variation is held constant over all elements. Consider a simple beam with parabolic shear loading as suggested by Robinson (1978). The error in the free end displacement v_1 is shown in Fig. 7 as the aspect ratio increases from 0.5 to 8. Only one element is considered, but the type, or order, is varied and indicated by their system name EPnn, nn denoting the number of degrees of freedom on the element. Thus, EP16 is the quadratic quadrilateral whereas EP24 is cubic. EP12 is the quadratic triangle, either 2 or 4 being used to model the given shape.

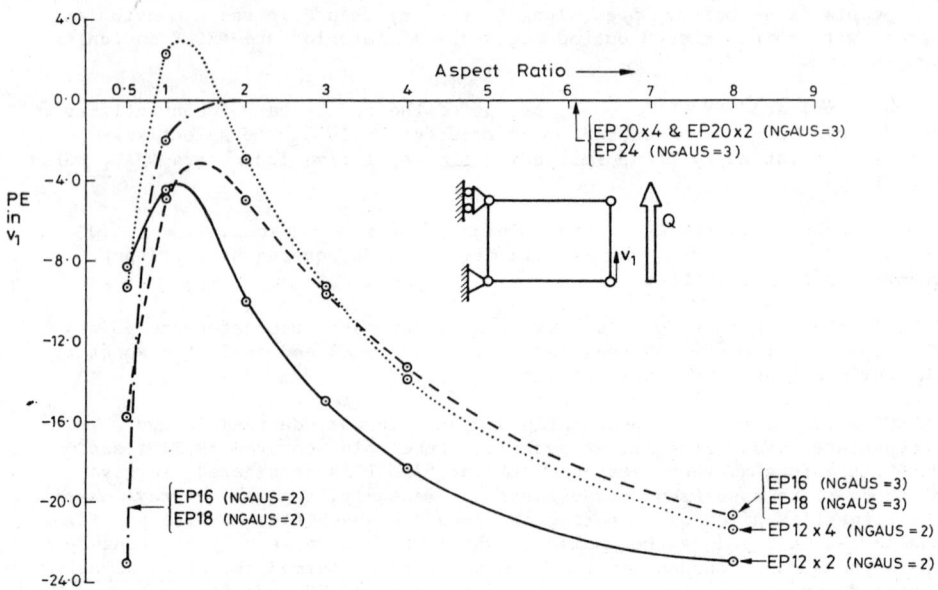

Fig. 7 Percentage of error in tip displacement v_1 (parabolic shear loading) single element test

Figure 7 demonstrates the error in v_1 increasing rapidly to over 20% in all cases except EP16 and EP18 (the lagrangian element) with NGAUS=2, and the cubic triangle EP20 and the cubic quadrilateral EP24, which remain accurate.

Since errors of high aspect ratios are due to implicit mesh coarseness, it would be expected that these errors decrease as numbers of elements increase. The above example has been repeated when, using the maximum aspect ratio of 8, meshes of 2, 4 and 8 elements along the lengths of the beam have been used whilst retaining the same aspect ratio. Figure 8 shows the rapid decrease in percentage error with increasing number of elements.

Thus, for real meshes where refinements have been carefully considered by the user, reasonable aspect ratios would not be expected to introduce serious errors, but nevertheless a careful scrutiny of the maximum aspect ratio per element should be made from the program's output. In particular, aspect ratios which exceed the input parameter ASPECT should be carefully checked.

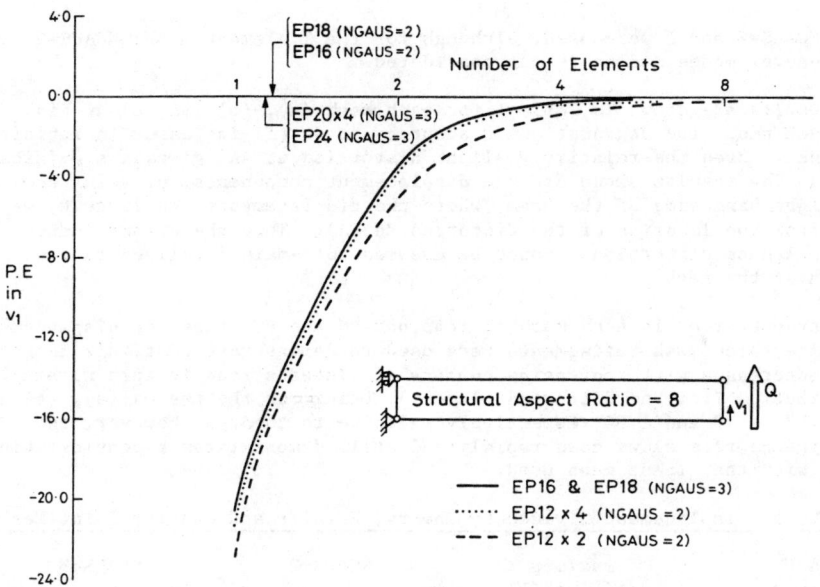

Fig. 8 Percentage of error in tip displacment v_1 (parabolic shear loading) One-dimensional convergence test

The Effect of Variations in det(J) on a Beam

The effects of variations in det(J) in computed results can be readily demonstrated in a simple beam. A square beam is loaded by parabolic shear only, and fixed as shown in Fig. 9(b).

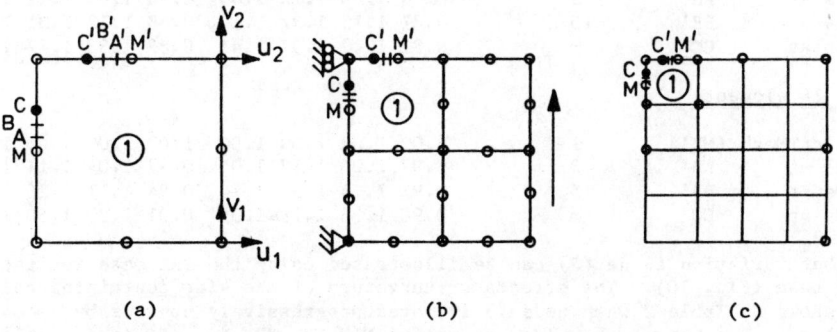

Fig. 9 Element meshes for beam

Three mesh refinements are considered, with 1, 4 and 16 quadratic displacement, 8 node elements in plane-stress. With the regular meshes as shown, excellent accuracy is apparent. However, to simulate a deterioration in det(J) in one element in each mesh, two midside nodes at the top left corner were progressed towards the quarter points adjacent to the top left hand vertex node, in equal steps as, initially, MM' (proper positions), AA', BB' then CC' (quarter positions), thus activating the checks against DISNOD.

Both NGAUS=2 and 3 were used, although for the 1 element mesh, NGAUS=2 gives zero energy modes and so is not considered.

The results are shown in Table 1 for each mesh (a), (b) and (c) of Fig. 9. For each mesh, the degradation in accuracy as det(J) increases to infinity is apparent. Even the relatively slight distortion at AA' gives a significant error. The results shown are the displacement components, u_1 v_1 u_2 and v_2 on the right hand side of the beam, where the displacements are largest, well away from the location of the distorted det(J). Thus the errors induced by local element distortions cannot be assumed to remain localised to that region of the mesh.

The largest error in each mesh corresponds to the CC' case for displacement u_2. The three mesh refinements were used to demonstrate that this largest error decreases with increasing numbers of elements (and in turn no smaller contribution from the distorted element), being for the meshes (a), (b) and (c) 2.57, 1.77 and 1.60 respectively relative to theory. However, the convergence rate slows down rapidly and still demonstrates a considerable error with the finest mesh used.

TABLE 1 Displacement Divided by Theory, Regular and Quarter Point Meshes

Mesh	max/min det(J) el(1)	NGAUS=2				NGAUS=3			
		u_1	v_1	u_2	v_2	u_1	v_1	u_2	v_2
(a) 1 element									
Regular Mesh (MM')	1			–		1.00	0.97	1.00	0.97
Nodes at AA'	3			–		0.86	1.01	1.21	1.03
Nodes at BB'	15			–		0.74	1.13	1.58	1.16
Nodes at CC'	∞			–		0.59	1.53	2.57	1.59
(b) 4 elements									
Regular Mesh (MM')	1	1.00	1.00	1.00	1.00	1.00	1.00	1.00	1.00
Nodes at AA'	3	0.94	1.05	1.13	1.05	0.94	1.04	1.12	1.04
Nodes at BB'	15	0.87	1.15	1.37	1.14	0.88	1.12	1.31	1.12
Nodes at CC'	∞	0.81	1.42	1.97	1.41	0.82	1.33	1.77	1.32
(c) 16 elements									
Regular Mesh (MM')	1	1.00	1.00	1.00	1.00	1.00	1.00	1.00	1.00
Nodes at AA'	3	0.97	1.05	1.12	1.05	0.97	1.04	1.11	1.04
Nodes at BB'	15	0.93	1.14	1.32	1.14	0.94	1.11	1.26	1.11
Nodes at CC'	∞	0.90	1.36	1.78	1.35	0.91	1.27	1.60	1.27

Another variation in det(J) can be illustrated using the 2x2 mesh for the same beam (Fig. 10). The effects of curvature of the side containing node 13 are shown in Table 2 when node 13 is moved progressively upwards by amounts equal to a quarter of the side, to points N1, N2, N3, N4. It is possible in this case to derive det(J) analytically as

$$\det(J) = a^2 - \frac{a}{2}(1 - \xi^2)k$$

where 2a = side length and ξ = 0 at node 13 and all points above it. k is the perturbation distance. In element 2, k is negative. det(J)max appears along lines ξ = 0 whilst det(J)min lies along lines ξ = ± 1, the left and

TABLE 2 Displacements Divided by Theory Regular and Curved Meshes

Mesh	max/min det(J) el(4)	max/min el(2)	NGAUS=2				NGAUS=3			
			u_1	v_1	u_2	v_2	u_1	v_1	u_2	v_2
Regular Mesh	1	1	1.00	1.00	1.00	1.00	1.00	1.00	1.00	1.00
Node 13 at N1	1.33	1.25	1.02	1.00	0.98	1.00	1.00	0.99	0.97	0.99
Node 13 at N2	2	1.5	1.06	1.01	0.94	1.01	0.97	0.97	0.91	0.97
Node 13 at N3	4	1.75	1.14	1.03	0.88	1.01	0.92	0.94	0.84	0.94
Node 13 at N4	∞	2	1.24	1.07	0.80	1.02	No solution			

right vertical sides. The results show the expected deterioration in accuracy as the det(J) ratio increases up to errors of 24% in u_1. The NGAUS=3 run with node 13 at N4 failed because det(J) was zero at 3 integrating points (all with $\xi = 0$) when calculating the element stiffness matrices. The effects of using reduced or complete integration produced roughly comparable errors although u_1 increases rather than decreases with the reduced rule.

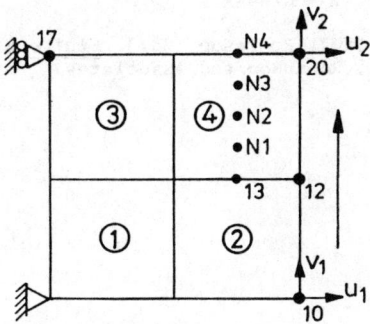

Fig. 10 2x2 Beam for Curvature Distortions

The two cases just illustrated both show a degradation in accuracy of key displacements as, somewhere in the mesh, det(J) approaches infinity. The first problem produces considerably larger errors than the second, even though the disturbed element is further away from the quoted displacements, and share only a small improvement with mesh refinement.

These results demonstrate that, although it is not possible to devise hard and fast rules on accuracy against det(J) variation within an element, nevertheless the user should be aware of such variations. In most practical meshes, det(J) max/min will not usually exceed 2, indeed as shown earlier big aspect ratios can exist with det(J) max/min equal to unity. The program output highlights such values which exceed the input parameter RATIOJ, which defaults to 2.

CONCLUSIONS

The BERQUAL program has been described for computerised quality assessments of finite element meshes. It fills an important gap between data generation and main analysis in finite element technology. No matter how correct the given data is, the effects of certain global or local parameters can have devastating effects on quality of results. It is not possible to devise hard and fast rules on how much the given parameters may distort, although the examples given show how trends for a given type of analysis could be established. With increased experience, it should be possible to categorise analyses into families for which the parameters calculated would only be accepted if within known tolerances, and possibly linked with expert systems.

ACKNOWLEDGEMENT

The work was carried out at the Berkeley Nuclear Laboratories of the Technology Planning and Research Division, and the paper is published by permission of the Central Electricity Generating Board.

REFERENCES

Hellen, T.K. (1976). In J.R. Whiteman (Ed.), The Mathematics of Finite Elements and Applications II, Academic Press, London. Chap. 40, pp 511-524.

Irons, B.M.R., and Ahmad, S. (1980). Techniques of Finite Elements. Ellis Horwood Ltd., Chichester, England.

Robinson, J. (1978). In J. Robinson (Ed.), Finite Element Methods in the Commerical Environment, Robinson and Associates. Paper 12.

A THERMAL SHELL FINITE ELEMENT FOR THE THERMOMECHANICAL ANALYSIS OF THIN SHELLS

J. L. Blanchard and E. Carnoy

Framatome/CS, la Boursidière, RN 186, B.P. 80, 92357 Le Plessis Robinson, France

ABSTRACT

This paper presents a thermal shell element for the computation of 3D temperature fields in structures modelled using isoparametric shell elements. The element provides through thickness temperature profiles with an accuracy comparable to those produced by more expensive types of modelling. The method employed involves representing the temperature profile along the fibers by a Legendre development. One to six polynomials are used.

KEYWORDS

Thermal shell, Legendre polynomials.

NOMENCLATURE

$\vec{X}$	Position of a point in physical space
$\vec{X}_i$	Position of node i in the reference surface
$\vec{N}_i$	Unit vector indicating the fiber direction at node i
h_i	Fiber thickness at node i along $\vec{N}_i$
(ξ,η,ζ)	Parametric coordinates
$N_i(\xi,\eta)$	Serendipity shape function associated with node i
n	Number of nodes
m_i	Legendre cardinal at node i
$P_j(\zeta)$	Legendre polynomial j
Θ_{ij}	Generalized temperature coordinate associated with node i and Legendre polynomial j.
T	Temperature (°C)
t	Time (s)

ρ	Density (kg m^{-3})
c_p	Thermal capacity (J kg^{-1} °C^{-1})
k	Thermal conductivity (Wm^{-1} °C^{-1})
[J]	Jacobian matrix
[Ĵ]	Inverse of matrix [J]
Ω	Finite element
dω	Differential volume
K	Notation or index relating to the conduction terms
Q	Notation or index relating to the source term
C	Notation or index relating to the thermal capacity terms
F	Notation or index relating to a Fourier condition
h	Heat exchange coefficient (Wm^{-2} °C^{-1})
Σ	Side of a finite element
dτ	Differential surface
N	Notation or index relating to a Neumann condition
q	Heat flux (Wm^{-2})
δt	Time step (s)

SYMBOLS

[ ]	Matrix
{ }	Column matrix
$[\ ]_p^r$	Matrix assessed at time step i and iteration p
—	For an imposed quantity
Δ	Laplace operator
^	Relating to the reference element

INTRODUCTION

Thermomechanical analysis is often based on complex 3D structures modelled by isoparametric shell elements. Certain fast breeder reactor components provide examples of the need for such an approach. These analyses necessitate the accurate determination of the thermal distribution within the component in order to obtain results comparable with those from a more expensive solid element computation.

The thermal shell element described in this paper achieves this accuracy by using a Legendre development to represent the temperature profile along the fibers, and conventional Serendip functions to approximate the thermal field accross the lamina. The thermal analysis uses the same mesh as the mechanical analysis.

The present paper describes the formulation of the main matrices and the approach to boundary conditions. The problem of branched-shell elements is also examined, with its implications regarding junction modelling. The paper also discusses the limitations inherent in the Legendre development with respect to the representation of some transient thermal profiles.

MATRIX FORMULATION

Geometrical description

The structure is modelled by a set of nodes arranged on a reference surface. Each node, characterized by a position vector $\vec{X}_i$, has an associated unit vector indicating the fiber direction, $\vec{N}_i$, and a fiber thickness h_i (fig.1). The thermal shell element uses the same geometry as the usual degenerated isoparametric shell element, with Serendipity shape functions along the lamina. Using the rotation N_i (ξ,η) to represent the Serendip functions related to each node, the transformation of the bi-unit cube (fig.2) into a physical space element is written :

$$\vec{X}(\xi,\eta,\zeta) = \sum_{i=1}^{n} N_i(\xi,\eta) \, (\vec{X}_i + \frac{1}{2} h_i \zeta \vec{N}_i) \qquad (1)$$

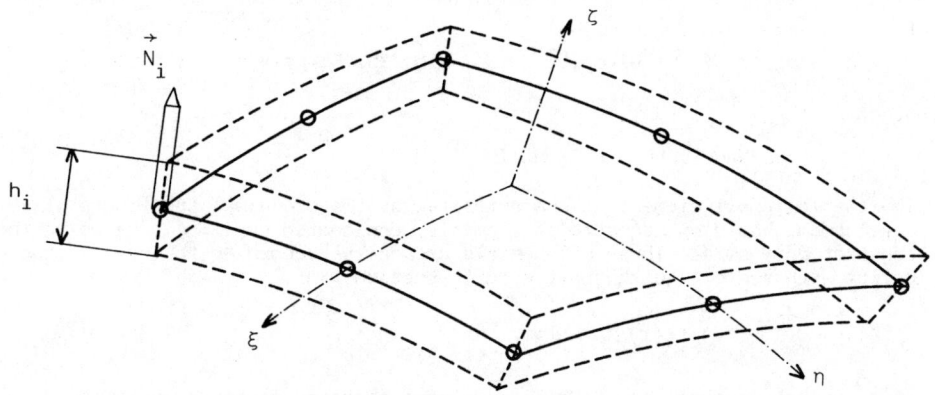

Fig. 1 - Thermal shell element in physical space

For ζ fixed, the surface defined by (1) is called a lamina. For (ξ,η) fixed, the line defined by (1) is called a fiber.

The method for determination of the unit vectors $\vec{N}_i$ between the fibers of adjacent elements is described in detail by CARNOY (1983). The principle is as follows :

- for each element, formation at each node of the normal to the reference surface ;

- where a node is shared by more than one element, computation of the matrix of scalar products of the different normals ;

- processing of the matrix to obtain characterization of the directions considered as independent with respect to a given criterion.

A branched-shell node is characterized by several independent directions. In a standard node, on the other hand, a unique fiber is defined by the combination of the different directions.

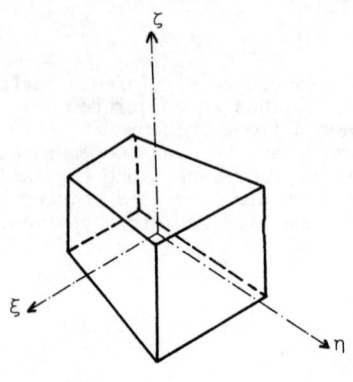

Fig. 2a Diagram of the bi-unit cube

Fig. 2b. Bi-unit cube : numbering.

THERMAL FIELD APPROXIMATION

The thermal shell element is characterized by the use along the fibers of a development based on Legendre polynomials, designated by $P_j(\zeta)$, P_1 being the constant polynomial. These polynomials are chosen according to their orthogonality with respect to the unit weight function :

$$\int_{-1}^{1} P_i(\zeta) P_j(\zeta) d\zeta = 0 \text{ si } i \neq j \qquad (2)$$

This property ensures compatibility between adjacent elements where the degree of through-thickness interpolation differs.

The generalized temperature coordinates are designated Θ_{ij}, where the first index indicates the node number and the second the index of the Legendre polynomial. The thermal field approximation of the reference element is written :

$$T(\xi,\eta,\zeta) = \sum_{i=1}^{n} \sum_{j=1}^{m_i} \Theta_{ij} N_i(\xi,\eta) P_j(\zeta) \qquad (3)$$

The number of nodes in the element, n, varies between four and eight, according to whether the shell sides are linear or parabolic. The number of generalized degrees of freedom m_i varies between one and six. The value two supplies, in particular, the usual linear profile of the thermal shell elements. For the branched-shell nodes, the Legendre cardinal is reduced to one. This means that the geometrical discontinuity corresponding to a branched-shell node prevents transmission of the thermal profiles. Only the first generalized coordinate, representing the mean temperature of the fiber, is transmitted.

A limitation of the element resides in the fact that a Legendre representation could be inadequate for certain thermal profiles. A Dirac impulse develops, for instance :

$$\delta(\zeta - 1) = \sum_{j=1}^{m} \frac{2j-1}{2} P_j(\zeta) \tag{4}$$

This development (4) is not convergent, since its coefficients are ascending. This explains the oscillatory behaviour of the first increments in a fast thermal transient calculation. In the first step, the thermal profile along the fibers, resembling a Dirac function, does not converge. In the second step, penetration of the thermal shock induces convergence. However, such a limitation concurs with the objectives of the element. For such transients, very fine through-thickness modelling of the structure is necessary.

DISCRETIZATION OF THE HEAT EQUATION

The discretization of the heat equation, expressed as follows :

$$\rho c_p \frac{\partial T}{\partial t} = k \Delta T + Q \tag{5}$$

is accomplished by minimization of the functionals as expressed by HUEBNER (1975). This minimization leads to the elementary matrix system :

$$\frac{\partial}{\partial t}[C]\{\Theta\} + [K]\{\Theta\} = \{Q\} \tag{6}$$

where $\{\Theta\}$ designates the generalized coordinates vector, the general term of which is Θ_{ij}. The thermal conductivity matrix [K] is expressed :

$$K_{ij,kl} = k \int_{\hat{\Omega}} (\alpha_{ij}\alpha_{kl} + \beta_{ij}\beta_{kl} + \gamma_{ij}\gamma_{kl}) |\det[J]| \, d\hat{\omega} \tag{7}$$

involving the following quantities

$$\begin{Bmatrix} \alpha_{kl} \\ \beta_{kl} \\ \gamma_{kl} \end{Bmatrix} = [\hat{J}] \begin{Bmatrix} P_l \, \partial N_k/\partial \xi \\ P_l \, \partial N_k/\partial \eta \\ N_k \, dP_l/d\zeta \end{Bmatrix} \tag{8}$$

The thermal capacity matrix [C] is written :

$$C_{ij,kl} = \rho c_p \int_{\hat{\Omega}} N_i(\xi,\eta) \, P_j(\zeta) N_k(\xi,\eta) P_l(\zeta) |\det[J]| \, d\hat{\omega} \tag{9}$$

Considering only uniform voluminal heat sources, the source vector $\{Q\}$ is :

$$Q_{ij} = Q \int_{\hat{\Omega}} N_i(\xi,\eta) \, P_j(\zeta) |\det[J]| d\hat{\omega} \tag{10}$$

BOUNDARY CONDITIONS

For the Dirichlet conditions, two cases must be distinguished. The first case corresponds to a temperature $\overline{T}(\zeta)$ imposed along a fiber. The development (3) and the orthogonality of the Legendre polynomials can then be used to obtain directly the generalized coordinates concerned. For example :

$$\Theta_{2k} = \frac{2k-1}{2} \int_{-1}^{1} \overline{T}(\xi=1, \eta=-1, \zeta) P_k(\zeta) d\zeta; \quad 1 \leq k \leq m_2 \tag{11}$$

The second case corresponds to a temperature imposed on the upper face ($\zeta = 1$) or the lower face ($\zeta = -1$) of the shell, normally to a node. Development (3) then provides a linear relationship between the generalized coordinates of the node concerned. For example :

$$\overline{T}(\xi=1, \eta=-1, \zeta=\pm 1) = \sum_{j=1}^{m_2} \Theta_{2j} P_j (\zeta=\pm 1) \tag{12}$$

For the Fourier conditions, defining exchanges with a fluid, only cases where these exchanges concern the upper or lower faces of the shell are considered. The fluid temperature and the heat exchange coefficient are presumed to be uniform. Discretization of the Fourier conditions requires minimization of the functionals described by HUEBNER. This leads to add to the second member of matrix system (6) the quantity :

$$\{F\} - [F] \{\Theta\} \tag{13}$$

The general term of the Fourier vector is :

$$F_{ij} = h_F T_F \int_{\hat{\Sigma}} N_i(\xi, \eta) P_j(\zeta = \pm 1) d\hat{\tau} \tag{14}$$

The Fourier matrix is written :

$$F_{ij,kl} = h_F \int_{\hat{\Sigma}} N_i(\xi, \eta) P_j(\zeta=\pm 1) N_k(\xi, \eta) P_l(\zeta=\pm 1) d\hat{\tau} \tag{15}$$

A Neumann condition, defining an imposed flux q_N, results in a vector $\{N\}$ similar to the Fourier vector :

$$N_{ij} = \overline{q}_N \int_{\hat{\Sigma}} N_i(\xi, \eta) P_j(\zeta=\pm 1) d\hat{\tau} \tag{16}$$

ASSEMBLY

Let the first element be Ω^1, where the cardinals of nodes i are m_i^1. The cardinal m^1 associated with the element is defined by :

$$m^1 = \max_i m_i^1 \tag{17}$$

Similarly, an element Ω^2 adjacent to Ω^1 has the cardinal m^2. Two assembly procedures are used, the first kinematic, the second static. The first procedure consists in writing for the nodes i of the common side :

$$\begin{array}{ll} \Theta_{ij}^1 = \Theta_{ij}^2 & j \leq \min(m_i^1, m_i^2) \\ \\ \Theta_{ij}^1 = \Theta_{ij}^2 = 0 & j > \min(m_i^1, m_i^2) \end{array} \tag{18}$$

The second consists in writing :

$$\Theta^1_{ij} = \Theta^2_{ij} \qquad j \leq \min(m^1, m^2)$$

$$q^1_{ij} = q^2_{ij} = 0 \qquad j > \min(m^1, m^2) \tag{19}$$

where q_{ij} is the heat flux associated with the generalized coordinate Θ_{ij}. In the first procedure, the generalized coordinates of a higher index than the m^1_i and m^2_i minimum are eliminated. In the second, the generalized coordinates of a higher index than the m^1_i and m^2_i minimum are condensed.

Computational strategies

The matrix equation resulting from the spatial discretization step is written :

$$\frac{\partial}{\partial t}[C]\{\Theta\} + [K]\{\Theta\} = \{Q\} + \{N\} + \{F\} - [F]\{\Theta\} \tag{20}$$

For the sake of simplicity, we write :

$$\{S\} = \{Q\} + \{N\} + \{F\} \tag{21}$$

Similarly, the Fourier matrix is added to the stiffness matrix. An implicit time scheme is used :

$$\left(\frac{1}{\delta t}[C] + [K]^{r+1}\right)\{\Theta\}^{r+1} = \{S\}^{r+1} + \frac{1}{\delta t}[C]\{\Theta\}^r \tag{22}$$

The thermal capacity matrix is assumed to be constant.

During linear analysis, the material properties required to form the quantities $[K]^{r+1}$ and $\{S\}^{r+1}$ are derived from the generalized coordinates of the previous increment. During non-linear analysis, on the other hand, the generalized coordinates of increment r+1 are computed by successive approximations. For this purpose, equation (22) is used as the recurrent relationship :

$$\left(\frac{1}{\delta t}[C] + [K]^{r+1}_p\right)\{\Theta\}^{r+1}_{p+1} = \{S\}^{r+1}_p + \frac{1}{\delta t}[C]\{\Theta\}^r \tag{23}$$

The material properties necessary for the formation of quantities $[K]^{r+1}_{p+1}$ and $\{S\}^{r+1}_{p+1}$ are, in this case, derived from the generalized coordinates of iteration p.

APPLICATIONS

The applications presented here were processed by the NOVNL code (CARNOY and PANOSYAN, 1985). In the near future, the thermal shell element is planned to be incorporated into the general purpose SYSTUS code (1986).

Steady-state temperature field in a T-junction

This case is intended to assess the effect of branched-shell nodes (fig.3). The layout of the boundary conditions is such that a significant portion of the imposed flux flows through the junction. Finite difference and thermal shell analyses have both been carried out (fig.4). The latter analysis uses both of the assembly procedures.

The same temperature map is obtained with both assembly procedures (fig.5), tending to indicate that convergence is achieved, if we accept, by analogy with the mechanical analysis, that these two procedures frame reality. The finite difference temperature map (fig.6) confirms this assumption. The difference between the results obtained with the two models is, as can be seen, very slight.

Cold shock test on tubes

This case analyzes the non-linear evolution of a temperature profile during a fast transient. It is taken from a computer code validation report, based on experimental tests (PANOSYAN, 1984). It concerns a thick, infinitely long tube at an initial temperature of 400 °C, which is subjected to the contact of a fluid on its outer surface.

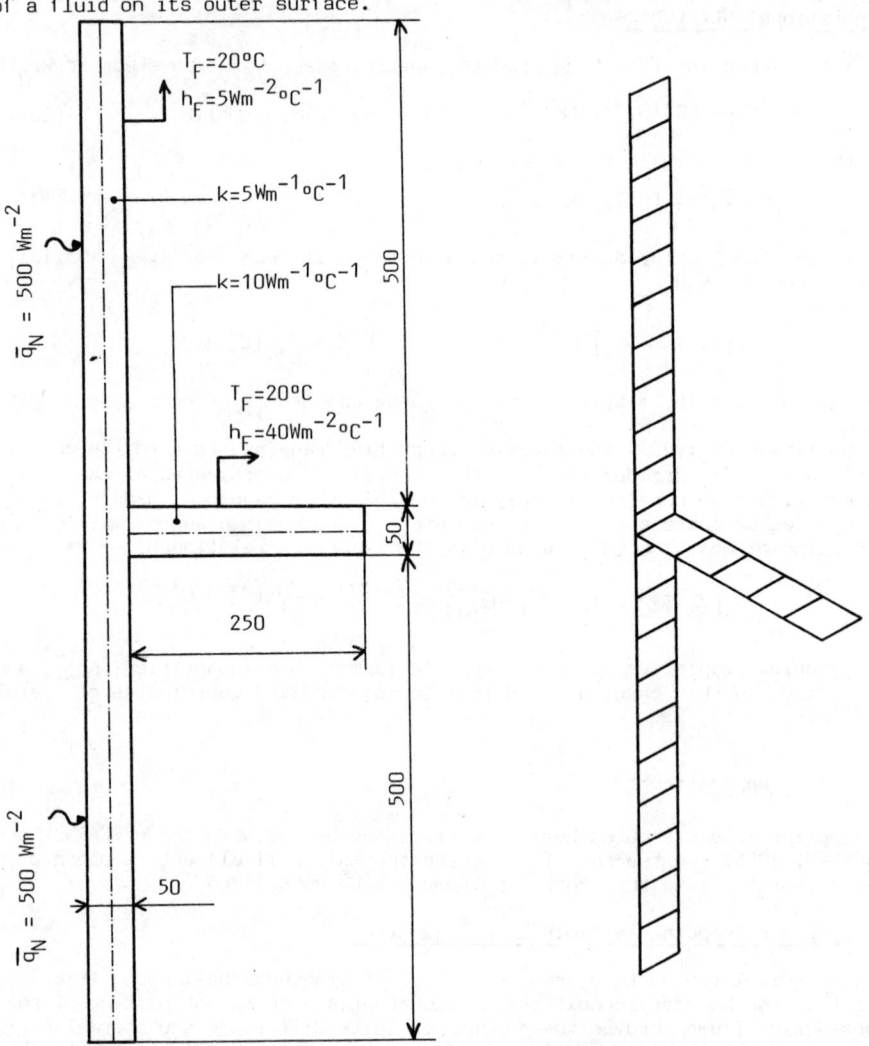

Fig. 3 Diagram of the T-junction   Fig. 4 T-junction : thermal shell element mesh

In the reference solution, the tube is modelled through its thickness by ten 2D axisymmetric elements. The present solution uses a single thermal shell element with five Legendre polynomials (fig.7). The time strategies are identical. It must be noted, however, that the product of the density by the specific heat retains its initial value throughout the present computation (table 1).

23	119	117	114						119	116	114					
22	119	116	114						119	116	114					
21	118	116	114						118	116	114					
20	118	115	113						118	115	113					
19	116	114	112						116	114	112					
18	115	112	110						115	112	110					
17	113	110	108						113	110	108					
16	110	108	105						110	108	105					
15	107	104	102						107	104	102					
14	102	100	98						103	100	97					
13	97	95	93	63	50	41	36	35	101	96	86	64	50	41	37	35
12			91	67	52	42	37	36	106	101	90	67	52	43	38	36
11	118	116	116	68	52	43	38	36	118	112	98	69	53	44	39	37
10	166	164	163						160	157	155					
9	208	206	205						201	199	199					
8	246	244	243						239	237	236					
7	278	276	276						271	269	269					
6	306	304	303						299	297	296					
5	328	326	326						321	319	319					
4	346	344	343						339	337	336					
3	358	356	356						351	349	349					
2	366	364	363						359	357	356					
1	368	366	366						361	359	359					
	1	2	3	4	5	6	7	8	1	2	3	4	5	6	7	8

Fig.5 T-junction : thermal shell element temperature map

Fig.6 T-Junction : finite difference temperature map

Table 1 - Cold shock test : material properties

Temperature (°C)	20	125	300	400
ρc_p product ($J°C^{-1}m^{-3}$)	$3,6 \cdot 10^6$	$3,95 \cdot 10^6$	$4,24 \cdot 10^6$	$4,31 \cdot 10^6$
Conductivity ($Wm^{-1}°C^{-1}$)	13,5	15,0	17,4	18,8

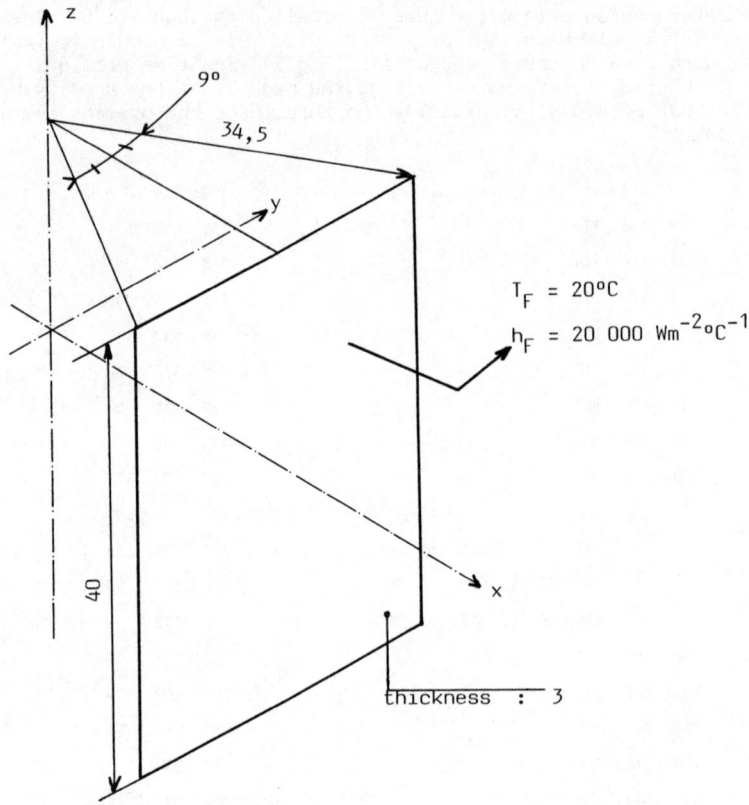

Fig. 7 - Diagram of the cold shock test

Table 2 summarizes the values obtained with the present solution during the first increments. At instant 0.001 second, the mean radius temperature exceeds that of the inner radius. At instant 0.04 second, the inner radius temperature exceeds the initial temperature. For the first increment, the ascending generalized coordinates reflect the non-convergence of the Legendre development related to the approximation of a Dirac function (4). At instant 0.1 second, on the other hand, the development becomes convergent.

Table 2 - Cold shock test : first increments

t (s)	Generalized coordinates					Temperatures (°C)		
	1	2	3	4	5	R_i	R_m	R_e
0,001	399,4	-1,7	-2,6	-3,7	-4,3	397,1	399,1	387,1
0,04	382,8	-39,6	-38,3	-27,0	-10,1	401,1	398,2	267,8
0,1	362,5	-71,7	-50,1	-17,6	-1,9	399,7	386,8	221,2

Table 3 summarizes the values obtained from the analytical model shown in figure 8 (LUIKOV, 1986). This model confirms that at instant 0.1 second, the thermal shock penetration reaches the inner radius of the tube.

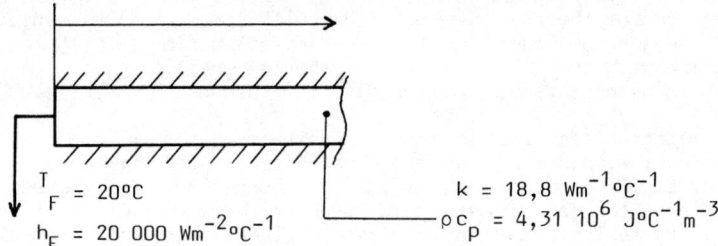

Fig. 8 - Cold shock test : analytical model

Table 3 - Cold shock test : analytical model

t(s)	Temperatures (°C)		
	R_i	R_m	R_e
0,001	400,00	400,00	371,65
0,04	400,00	399,28	265,24
0,1	399,89	387,51	219,47

From table 4 it can be seen that, in spite of the simplicity of the model, there is an excellent correlation between the results of the thermal shell analysis and the reference solution.

table 4 : Cold shock test : comparison between the reference solution (RS) and the thermal shell (TS)

t(s)	inner radius		outer radius		t(s)	inner radius		outer radius	
	RS (°C)	TS (°C)	RS (°C)	TS (°C)		RS (°C)	TS (°C)	RS (°C)	TS (°C)
0,1	399	400	218	221	0,2	393	393	178	182
0,3	379	379	157	160	0,4	362	361	142	144
0,5	344	343	130	132	0,6	325	324	121	123
0,7	307	306	113	114	0,8	290	289	106	107
0,9	274	272	100	100	1,0	258	256	94	95

CONCLUSION

The practical cases described show that the level of accuracy obtained by expanding the temperature profile along the fibers in a Legendre development is adequate for the requirements of a shell type modelling method. In particular, the non-convergence of this development during fast thermal transients is in keeping with the objectives of the thermal shell element, since the analysis of such transients requires extremely fine through-thickness modelling.

This limitation also implies that excessive reduction of the time step would be ineffective, since an adequate penetration of the thermal shock is necessary for the development to converge. The convergence can be easily verified by examining the generalized coordinates. Advantage could be taken of this possibility to introduce surface temperature corrections, eventually necessary for the mechanical analysis, by means of a 2D conventional discretization of the thermal profile along the fiber.

The use of the same mesh for the thermal and mechanical analysis constitutes a considerable advantage of the thermal shell element, since the user can apply the thermal loads directly to the mechanical mesh. The computational processing drawbacks associated with the conventional interpolation procedures are thereby avoided.

REFERENCES

Carnoy E. (1983). Shell finite elements in ISPRA courses in structural dynamics. Commission of the European Communities (Ed.), Ispra, Italy.
Carnoy E. and Panosyan G., (1985). NOVNL CODE. NOVATOME technical report TAD85004.
Huebner, K.H. (1975). The finite Element Method for Engineers. Wiley Interscience. Chap.4, pp. 113-119.
Luikov, A.V. (1968). Analytical Heat Diffusion Theory. Hartnet, J.P. (Ed.), Academic Press. Chap. 6, pp. 203-205.
Panosyan G. (1984). Inelastic Benchmark Calculation.- Cold Shock Problem on AISI 316 SS tubes. NOVATOME Technical report MD84469, EEC, Working Group of Codes and Standards.
SYSTUS SYSTEM User's Manual (1986). FRAMATOME/CS.

ON A METHOD OF STRUCTURAL ANALYSIS

Guo Youzhong

Wuhan Institute of Mathematical Science, Academia Sinica, POB 30, Wuhan 430071, People's Republic of China

Abstract

In this paper we present the theoretical foundation concerning a famous method in structural analysis, the distribution of moment(or/and displacement): the mathematical description, new criterion, convergence and error estimation of approximate solution and some possible generalizations to that of other constructions.

§0. Introduction

The distribution of moment(or/and displacement)is a famous method in structural analysis, it is an iterative process in solving the static underdetermined problems.

As soon as Professor Hardy Cross of the United States had advanced the distribution of moment in 1930, the method was widely adopted by the international engineering and mechanical community. In the 50's, a great number of mathematicians, mechanicians and engineers were engaged in its generalization and application, to whom the scholars of our country also contributed so much.

In 1950, Professor C. V. Klauček of Czechoslovakia advanced the distribution of deformation from another point of view, it is a simplified method in view of the superposition principle, which is different from the distribution of displacement mentioned in this paper. Those two methods each having its strong points had wun universal praise.

Half century had passed, the distribution of moment was placed on the university courses supplying lots of basic knowledge and exercises, however, its strict mathematical foundation, convergence and error estimation had been out of sight, so that there had not been exact description of the essentials of the method and its applicable conditions. The main purpose of this paper is to make up this gap and solve the unsettled problem. Once the essence of the problem is revealed, further generalizations could become obvious.

§1. Displacement Method and Distribution of Moment

The distribution of moment is an iterative solution of the displacement method.

1. In solving static underdetermined structures by means of the displacement method, the first thing to do is to define the unknown numbers of angular displacement n_a and those of linear displacement of the nodes in the structure n_l, which are referred displacement numbers $n(\triangleq n_a + n_l)$. Then add n_a extra rigid-arms and n_l extra

supporting rods to prevent angular and linear displacements, which are referred to n extra connections. That decomposed the original structure into a series of known single span static underdetermined (or determined) rods, the so-called basic structure. And for n extra connections of the basic structure, we can obtain a system of simultaneous equations of displacement by means of static equilibrium conditions as follows:

$$AX = B, \qquad (1.1)$$

where

$$A \triangleq [A_{ij}]$$

is rigidity matrix of order n × n, and A_{ij} is the force (or moment) produced in node i, by the unit displacement $X_j = 1$ acting on node j; and

$$X \triangleq [X_i]$$

is displacement colume matrix (vector) of order n, and X_i is the displacement acting on node i; whereas

$$B \triangleq [B_j]$$

is force colume matrix (vector) of order n, and B_j is the force produced in node j under the action of load.

After a unit load having applied at the end k of a node ij, k=i or j, it is easy to give the figures of moment $\overline{M}_k$, shearing force $\overline{Q}_k$ and axial force $\overline{N}_k$ of the basic structure, and find the matrix elements A_{ij} from

the equilibrium condition or the section method. Under
certain condition, from the reciprocal principle of
flexibility and rigidity

$$-A_{ij} = \int \frac{\overline{M}_i M_j}{EJ} ds + k \int \frac{\overline{Q}_i Q_j}{GF} ds + \int \frac{\overline{N}_i N_j}{EF} ds, \qquad (1.2)$$

where E, G, F, J and k are respectively rod elastical
modulus, shear modulus, section area, rotating inertia
and distribution coefficient of shear stress, and
$[-A_{ij}] \triangleq -A$ is the negative of A. Equation (1. 2) is
called Maxwell-Mohr formula. Under the action of load P,
it is easy to find matrix elements B_i when the figures of
moment M_p, shear force Q_p and axial force N_p are given
so that

$$-B_i = \int \frac{\overline{M}_i M_p}{EJ} ds + k \int \frac{\overline{Q}_i Q_p}{GF} ds + \int \frac{\overline{N}_i N_p}{EF} ds. \qquad (1.3)$$

It is not difficult to prove the uniqueness and
existence of the solution of Eq. (1. 1)(Cf./2/). Having
found X, we can smoothly find the moment, shearing and
axial forces of this static underdetermined structure.
But when the structure is complex and sophisticated and
the unknown displacement number n are tremendous, it
turns rather difficult to solve Eq. (1. 1) manually, even
the computer could meet the problem of data explosion,
so that various approximate methods appear.

2. In the works of distribution of moment, it is point out that the method is applicable to certain approximate solutions of static undetermined structure in the cases of $n_1=0$ and varied section area. The Cross' method presented the directly perceived distribution mechanism of mechanics, which is convenient for the operation on the calculation sketch, and gives exactly the geometrical and physical intuition. This shows the extraordinary ability of generalization of the discoverer, and as above mentioned, it could find its practical value as yet. However, its mathematical foundation is deeply concealed.

The Cross' method is beginning at analysis of basic structure consisting of single span rods. Let's study any rod ab with varied section area.

When end b(a) is fixed, $S_a(S_b)$ denote the needed moment producing unit angular displacement $\varphi_a=1 (\varphi_b=1)$ at end a(b), called the rigidity at end a(b) of the rod; $C_{ab}(C_{ba})$ denote the ratio of moments $M_a^{(\cdot)}$ ($M_b^{(\cdot)}$) and $M_a(M_b)$ transmitting from end a(b) to b(a).

When end a produces angular displacement φ_a, and end b produces angular displacement φ_b and linear displacement

$\triangle(\triangleq \varphi L)$, the total moments of both ends a and b are respectively:

$$M_{ab} = S_a \varphi_a + C_{ab}\varphi_b - (1+C_{ab})\varphi;$$
$$M_{ba} = S_b \varphi_b + C_{ba}\varphi_a - (1+C_{ba})\varphi. \quad (1.\ 4)$$

In the isosection rod with length L, $S_a = S_b = 4EJ/L$, $C_{ab} = C_{ba} = \frac{1}{2}$, the above equations become

$$M_{ab} = (2EJ/L)(2\varphi_a + \varphi_b - 3\varphi);$$
$$M_{ba} = (2EJ/L)(2\varphi_b + \varphi_a - 3\varphi). \quad (1.\ 5)$$

In different supporting cases we can get different equations.

When there are η ($1 \leq \eta \leq n$) rods rigidly connected at b, the parameters belonging to rod ab are noted by the index (a) with parentheses, such as $S_b^{(a)}$ etc. Therefore, the total rigidity of node a is

$$A_{bb} = \sum_{a=1} S_b^{(a)} = \sum_{a=1} A_{bb}^{(a)}, \quad (1.\ 6)$$

and load P produces total node force

$$B_b = \sum_{a=1} B_{b(a)} = -\sum_{a=1} M_{b(a)} = -M_b, \quad (1.\ 7)$$

where $M_{b(a)}$ is the end moment produced by the load at rod ba, M_b is called nonequilibrium moment; and the ratio

$$\mu_b^{(a)} \triangleq A_{bb}^{(a)} / A_{bb}$$

is called the distribution coefficient of rod ba, therefore

$$M_{a(b)} = \mu_a^{(b)} M_a = -\mu_a^{(b)} B_a. \qquad (1.7)*$$

On the basis of the basic structure, let $\alpha=1,\ldots,\eta$, $\eta=\eta(a)$, $a=1,\ldots,n$ the method of moment distribution carries on the following iterative operation:

i) calculate $B_\alpha^{(a)}$, $S_\alpha^{(a)}$, $C_{\alpha a}$ and $\mu_\alpha^{(a)}$.

ii) according to certain order, e.e. $1,\ldots,n$, first, relax rigid arm 1, using (1.6), calculate B_1; and then using (1.7), distribute nonequilibrium moment $M_1^{[1]} \triangleq M_1$, we get $M_{1(a)}$; after transmission according to (1.8), we get $M_{a(1)}^{[1]}$; then fix rigid arm 1 again. Relax rigid arm 2 and repeat the above operation till the n-th rigid arm, which we call a circulation.

iii) in the k-th(k>1) circulation, the nonequilibrium moment of rigid arm a becomes only transmission moment

$$M_a^{[k]} \triangleq \sum_{\alpha<a} M_{\alpha(a)}^{[k]} + \sum_{\beta>a} M_{\beta(a)}^{[k-1]}, \qquad (1.9)$$

in accordance with (1.7) distribute $M_a^{[k]}$; and by (1.8) carry on transmission, we obtain $M_{\alpha(a)}^{[k]}$, $k=1,\ldots,n$.

iv) when the circulation comes to the m-th time, the nonequilibrium moment $M_a^{[m]}$ is no longer larger than the predetermined allowable error $\varepsilon>0$, i.e. when

$$|M_a^{[m]}| \leq \varepsilon, \qquad (1.10)$$

the operation finishes. Simultaneously, we obtain the moment of rod ab at end a of the rigid-arm

$$M_a = B_{a(b)} + \sum_{k=1}^{m} M_{a(b)}^{[k]}. \qquad (1.11)$$

The operations in the sletch and chart of the method, cf. e.g. /1/.

3. Here we are going to explicit the relation between the displacement method and the method of moment distribution.

Now we again explain Eq. (1.1) in a somewhat different way. First let's see the diagonal term, namely the nonequilibrium moment:

$$A_{jj}X_j = \sum_{\alpha=1}^{n} A_{jj}^{(\alpha)}X_j = \sum_{\alpha=1}^{n} \mu_j^{(\alpha)} S_j^{(\alpha)} X_j = \sum_{\alpha=1}^{n} M_{j(\alpha)} = M_j; \qquad (1.12)$$

then the non-diagonal term:

$$A_{ij}X_j = C_{ij} S_j^{(i)} X_j = M_{i(j)}, \qquad (1.13)$$

that is, the moment distribution method is, essentially, an iterative method of finding approximate solution $X^{[m]}$:

$$A_{ii}X_i^{[m]} = B_i - \sum_{j=1}^{i-1} A_{ij}X_j^{[m]} - \sum_{j=i+1}^{n} A_{ij}X_j^{[m-1]}, \quad i=1,\ldots,n, \qquad (1.14)$$

when $m=1$, $X_j^{[m-1]}$ should be known to be naught. In practical operation, the following iterative formula is adopted:

$$A_{ij}(X_i^{[m]} - X_i^{[m-1]}) = -\sum_{j=1}^{i-1} A_{ij}(X_j^{[m]} - X_i^{[m-1]})$$
$$+ \sum_{j=i+1}^{n} A_{ij}(X_j^{[m-1]} - X_j^{[m-2]}) = -\sum_{\alpha<i} A_{i\alpha}(X_\alpha^{[m]} - X_\alpha^{[m-1]})$$
$$+ \sum_{\alpha>i} A_i(X_\alpha^{[m-1]} - X_\alpha^{[m-2]}), \quad m \geq 2. \qquad (1.14)^*$$

Let

$$H \triangleq I - \text{diag}(A_{ii}^{-1})A,$$
$$G \triangleq \text{diag}(A_{ii}^{-1})B, \qquad (1.15)$$

decompose H into the sum of the strictly lower trangular matrix L and strictly upper triangular matrix U, the equivalent matrix from Eq. (2.14) is

$$X^{[m]} = LX^{[m]} + UX^{[m-1]} + G, \qquad (1.16)$$

or

$$X^{[m]} = (I-L)^{-1}UX^{[m-1]} + (I-L)^{-1}G. \qquad (1.16)*$$

From Eq. (1.16) or (1.16)* one can really state that the method of moment distribution and Gauss-Seidel iteration method for the displacement method are just the same to each other!

§2. The Convergence and Error Estimation of Distribution of Moment

1. Let us consider Eq. (1.1).

From Eq. (1.2) we know:

$$A^-_{ij} = A^-_{ji}, \qquad (2.1)$$

this is the equivalent principle of displacements, therefore A^{-1} is real symmetric matrix; and

$$A^-_{ii} > 0. \qquad (2.2)$$

As for the rods with equi-section, generally, we have

$$A^-_{ii}(a) + A^-_{ij}(a) \geq 2|A^-_{ij}(a)|,$$

but $A^-_{ii}(a) = A^-_{ij}(a)$, hence $A^-_{ii}(a) \geq |A^-_{ij}(a)|$, only when $M_i^{(a)} = M_j^{(a)}$ the equal sign holds. In other cases, we can obtain applicable condition by use of this method. Since when η rods rigidly connected at end i are not all the latter, i.e.

$$A^-_{ii} = \sum_{\alpha=1}^{\eta} A^-_{ii}(\alpha) > \sum_{\alpha=1}^{\eta}|A^-_{\alpha i}| = \sum_{\alpha \neq i}|A^-_{\alpha i}|. \qquad (2.3)$$

Hence all A^{-1}'s principal minors $A^-_i > 0$ and A^{-1} is positively definite matrix, and so is A. From (1.5-1.13), we know, in fact, that in rod ij, when end i and j are fixed, $A_{ji} = ½$; when end j is supported with a hinge, $A_{ji} = 0$; but when end j is sliding-fixed, $A_{ji} = -1$.

Because A is a symmetric positively definite matrixes, Gauss-Seidel iterative method converges. Not only so, that Jacobi iterative method

$$X^{[m]} = HX^{[m-1]} + G, \quad m = 1,2,\ldots, \qquad (2.4)$$

and the over relaxation iterative method

$$X^{[m]} = (1-\omega)X^{[m-1]} + \omega(LX^{[m]} + UX^{[m-1]} + G),$$
$$0 < \omega < 2, \quad m = 1, 2,\ldots, \qquad (2.5)$$

are all convergent.

2. From (2.3) we can see that

$$\|H\| \triangleq \max_i \sum_{j=1}^{n} |H_{ij}| = \max_i \sum_{i \neq j} |A_{ij}|/A_{ii} < 1, \qquad (2.6)$$

write

$$\mu \triangleq \max_i \left[\sum_{j=1}^{n} |H_{ij}| / (1 - \sum_{j=1}^{i-1} |H_{ij}|) \right], \qquad (2.7)$$

we have the error estimation

$$\|X^{[m]} - X^*\|_\infty \leq \mu^m(1-\mu)^{-1} \|X^{[2]} - X^{[1]}\|_\infty, \qquad (2.8)$$

where X^* is the exact solution of Eq. (2.1), or

$$\|X^{[m]} - X^*\|_\infty \leq \mu(1-\mu)^{-1} \|X^{[m]} - X^{[m-1]}\|_\infty. \qquad (2.9)$$

From (2.9) we see that for $\mu \ll 1$, the quantity $\|X^{[m]} - X^{[m-1]}\|_\infty$ could be used to determine whether the iterative process should be stopped.

Up till now we have explicited the mathematical description, convergent condition, error estimation and the reasonableness for (1.10) regarded as the criterion of stopping the iterative operation of the moment distribution method.

§3. **A New Criterion of Convergence in Moment Distribution**

In the proceding section we have proved the equivalence between moment distribution and Gauss-Seidel method, then write $S(A)$ as the spectral radius of matrix A, immediately we know that the sufficient and necessary condition in moment distribution is also $S(1-L)^{-1}U) < 1$. Since this condition is not convenient for checking, therefore, people have to look for some new criterion in another way.

As a matter of fact, the essentials of convergence in all iterative methods are the contractibility of mapping, and what is different of moment distribution from other methods is that before calculating the component of $X_i^{[m]}$, each $X_j^{[m]}$, $j<i$, is already calculated so that it participates contraction a little former (Cf. the first term of Eq. (3. 1)). Exactly estimating the component participating contraction a little former in time, we have the following criterion:

Suppose that there is permutation $T \triangleq \begin{pmatrix} 1, \ldots, n \\ t_1, \ldots, t_n \end{pmatrix}$ and a group of positive numbers $\mu_k < 1$, such that

$$\sum_{s=1}^{k-1} |A_{t_k t_s} + \delta_{t_k t_s}|\mu_s + \sum_{s=k}^{n} |A_{t_k t_s} + \delta_{t_k t_s}| \leq \mu_k, \quad (3.1)$$

where $s,k = 1,\ldots,n$, $\delta_{t_k t_s}$ is Kronecker symbol. Then after permutating T, Eq. (1.16) becomes

$$X_{i_k}^{[m]} = \sum_{s=1}^{k-1}(A_{i_k i_s} + \delta_{i_k i_s}) X_{i_s}^{[m]} + \sum_{s=k}^{n}(A_{i_k i_s} + \delta_{i_k i_s})X_{i_s}^{[m-1]} + B_{i_k}$$

$$(3.2)$$

convergent, and also there is the estimate

$$|X_{ij}^m - X_{ij}^*| \leq \frac{|\mu|^m}{1-|\mu|} |X_{ij}^{[1]} - X_{ij}^{[0]}|, \qquad (3.3)$$

where

$$|f(j)| \triangleq \{\max f: 1 \leq j \leq n\}.$$

This means that although three norms of matrix (A+I) are all larger than 1, yet this method still converges. Thus the area of convergence for the moment distribution enlarged obviously, and the velocity of convergence increased greatly. It is only necessary to follow Eq. (3. 1) with the sequence (T) in selecting moment distribution.

§4. Force Method and Distribution of Displacement

When calculating the static underdetermined structure by force method, the form of (1. 1) is not changed at all, where n is the surplus constraint number of the static underdetrmined structure, after eliminating the surplus constraints and adding the unknown constraint force X_i, i=,,..,n, the basic structure is a static underdetermined structure: X is force vector acting on the surplus constraints; A is flexibility matrix under the acting of unit force; and B is displacement vector under the action of load. The difference between the two methods is vast in physics, but small in mathematics. At this very moment, in (2. 2) the negative index of $-A_{ij}$ should be omitted. Therefore, A is still a symmetric (the displacement reciprocal principle) positively definite matrix under certain cases, also there should be the corresponding distribution of displacement, thus the similar conclusions in §2 hold.

Similarly, for the mixed method, there should be a mixed distribution method.

It is especially in the 3-dimensional structures, such as in the net shell and some finite element and boundary element methods of the problems, after discretion, those problems can still be regarded as 3-dimensional rod structures. And the distribution of moment is also a good intuitive iterative method in certain cases as mentioned in the above section.

References

/1/ Jin Baozhen, Yang Shide and Zhu Baohua, Structural Mechanics, The People's Education Publishing House, 1964.

/2/ Guo Youzhong, Complementary variational principles in elastic theory, Kexue Tongbao, 29, 10, 1297-1302.

FINITE ELEMENT ANALYSIS OF FATIGUE CRACK GROWTH PROCESS

A. Bia and G. Pluvinage

Université de Metz, 57045 Metz, France

ABSTRACT

This paper presents an original and inexpensive finite element method of calculation for the analysis of any type of fatigue crack propagation, under any loading case.
A example is used to compare the simulation by the method of finite elements with experimental results, in the case which concerns the propagation of a fatigue crack under a loading of variable amplitude.

KEYWORDS

Fatigue; crack propagation ; simulation; finite elements; complex loading.

INTRODUCTION

All structural designers must, at some time or another, demonstrate the harmlessness of defects which are considered as cracks, in products for which they are responsible, whether they be discovered during manufacture, or by inspection in service
After metallurgical examinations, and destructive and non-destructive testing, comes the mechanical analysis , the collation of all the information, in order to decide upon the action to be taken:
can it be allowed to remain in its present condition ?
for how long ?
what is to be the frequency of the checks to monitor the crack in service ?
or must there be an immediate repair ?
Thus, in order to be sure of integrity of a structure, it is necessary to predict its life , taking into account the fact that it may contain defects such as fatigue cracks.
The calculation of the fatigue life of a structure which is subjected to random loadings is not a simple matter, due to

the continuous existence of transient phenomena,resulting
from the progress of the fatigue crack under actual loadings.
The problem of complex loadings is generally recorded on
structural elements as significant stresses,i.e. a loading
which is considered to be equivalent (as,for example ,in
the case of the method of programmed blocks).However,the
study of random fatigue is currently meeting with difficulties caused by the representativeness of these equivalent
loadings,the complexity,and the cost of running tests.

An initial approach to the problem consists of considering
a simple case,for example the application of a single overload (in producing a model of this problem it is assumed
that the effect of several overload peaks is deduced from
the effect of one overload).It is known that such a loading resuls in the slowing down of the progress of the cracks
and that one of the probable causes is the size of the residual stresses which develop at the tip of the crack.

The present work involves the numerical simulation of this
problem of fatigue cracking:
It is proposed to use a calculation code based on the method
of finite elements.An example has been used to compare the
simulation by the method of finite elements with experimental results,in the case of the propagation of a fatigue
crack under a loading of varying amplitude.

DESCRIPTION OF THE CODE

With the program SIFMEPA (SImulation of Fatigue and MEchanical ProgrAm) it is possible at the moment to deal with
plane elastic-plastic problems (plane stresses and plane
strains),as well as axisymmetrical ones.It is based on a
mixed method "IS.VS-CbC"-the initial stress method combined
with the variable stiffness method (Initial Stress ,
Zienkiewicz et al in 1969 ,Variable Stiffness ,Marcal et
al in 1967), and the changing boundary conditions method
(Changing boundary Conditions ,Kanninen in 1973,Newman in
1977 and Kaiser et al in 1983). Convergence (within the
constraints of the laws of plasticity)is ensured by the
initial stress method and accelerated by the change of
stiffness.The program is written in FORTRAN.

This code is designed to deal with problems concerned with
the mechanics of continuous media and the mechanics of the
fracture,in bidimensional fields,as well as problems concerned with welded joints.

STRUCTURE OF THE PROGRAM

The code consists of a set of modules for the static analysis of bidimensional structures by the method of finite
elements,as well as an automatic meshing module and a design module.The results of each module can be stored on file
to serve as data for other modules.

The structures are subjected to mechanical and thermal loadings which vary with time.
The laws to be taken into account, controlling the behaviour of the structures, are of the type:
+ elastic behaviour
+ elastic-plastic behaviour with or without work-hardening.

DESCRIPTION OF EACH MODULE

1-Meshing module AM (Automatic Meshing)

This module divides a plane or axisymmetrical part into elements which are:
+ standard triangular (3 nodes) or isoparametric (6)
+ standard rectangular (4 nodes) or isoparametric (8).

The division is determined from geometrical data for the contour of the part and the intended dimensions of the triangles/rectangles, and the numbering of the nodes in the final mesh is optimised by the program so as to minimise the processing time for the calculation by finite elements (Frederic et al in 1970, Lawrence et al in 1979), (Fig.1).

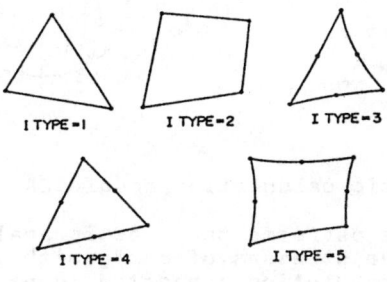

figure 1.

2-Design module IG (Interactive Graphics)

Both for checking the geometry of the mesh which is generated and for using the results, this module provides a graphic display of the distoritition of the part being studied and of the distribution of the values of the characteristic parameters within the part, (displacements , stresses and strains) either in digital form or in the form of curves, together with the results for the fatigue life and the rates of crack propagation (example , figure 2).

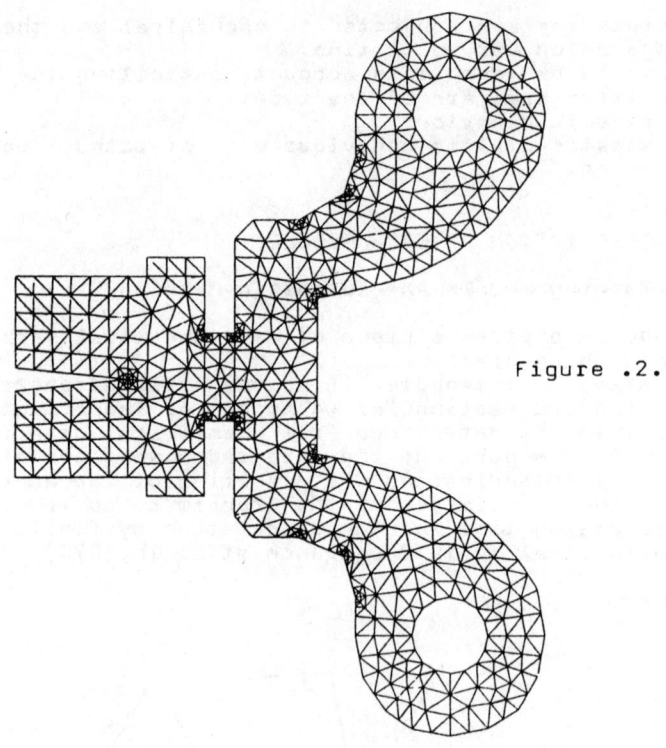

Figure .2.

3-Elastic calculation module EA (Elastic Analysis)

This module performs the elastic analysis of parts subjected to plane stresses,of parts with plane strains,and of solids of revolution subjected to axisymmetrical loadings, in terms of small strains and small displacements.The parts analysed can be composed of one or more zones of isotropic material.

The data required is as follows:
 -Geometry of the part (generated automatically by AM or defined by cards).If the geometry is defined by cards, the generation of the nodes and the elements is semi-automatic.Several nodes or elements can be generated by a single card.In addition ,the co-ordinates of the nodes can be given in cartesian or cylindrical form,in local bases.
 -Characteristics of each zone:
  + type of analysis (plane strains or stresses,axisymmetrical problem);
  + thickness of the zone (in the case of plane strains)
  + characteristics of the material (Young's Modulus, Poisson's ratio,density in the case of loading per

volume,coefficient of linear expansion in the case
of thermal loading).
-Displacement conditions :
   + imposed displacements of nodes or lines;
   + elastic supports at a node or a line;
   + perfect adherence,perfect sliding,or Coulomb friction,between the different portions of the part.
-Loadings :
Several loading cases can be dealt with simultaneously,
provided that the field which involves the conditions
affecting the displacements remains the same throughout the calculation.
Each loading case can contain loadings of the following types:
   + force concentrated at a node;
   + pressure and tangential load which are constant on
     a portion of the contour of a zone,or which vary
     linearly on one elemental dimension;
   + centrifugal force (rotation about the axes of the
     bidimensional model);
   + gravitational force;
   + thermally-induced loading.
-The results:
   The results obtained are the displacements,the strains
   the stresses,the principal stresses in intensity and
   direction,and the equivalent Von Mises and Tresca stresses for all the points in the mesh and at the integration points for each element.

4-Elastic-plastic calculation module EPA (Elastic-Plastic Analysis)

This module is used to resolve elastic-plastic problems(the
problem of cyclic plasticity) concerning two-dimensional
plane or axisymmetrical structures,using the method of
finite elements. This is done using the assumption "small
strains and large displacements".
The structures are subjected to mechanical and thermal loading,the characteristics of the materials,which are isotropic
or anisotropic,are a function of the temperature.
The laws to be taken into account,controlling the behaviour
of the structures,are of the type :
   +elastic behaviour
   +elastic-plastic behaviour with or without work-hardening (isotropic,Kinematic or mixed isotropic-Kinematic work hardening).
The materials are assumed to be elastic-plastic with mixed
isotropic-kinematic work-hardening (or isotropic work-hardening,kinematic work-hardening)-Thomas et al in 1983. The
structure may be the seat of a field of initial plastic
strains.

The data required is as follows:
   - Geometry of the part (generated automatically by AM
     or defined by cards);
   - The behaviour laws;

- The displacement conditions (imposed displacements and elastic supports);
- The loadings (the same as for EA, together with their variation with time).
-The results:
The program then determines the corresponding strain, stress and unelastic displacement fields, by an elastic calculation!
For the cyclic plastic analysis, it is assumed for the time being that the loading is radial. Thus data which corresponds to firstly the minimum and then the maximum loading condition must be introduced.
The results obtained are the same as those for EA, with in addition the plastic strain, for each step of the calculation demanded.

Remark:
+ run the program, declaring an elastic-plastic phase, with a single time step covering the whole of phase 1 and a single iteration to obtainn convergence (figure 3a);
+ restart at the end of the first phase, to analyse the second phase, and so on (figure 3b).

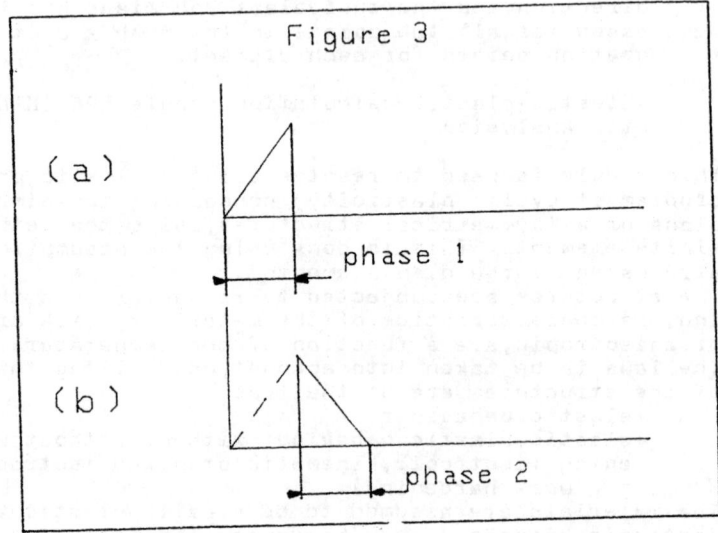

Figure 3

(a)

phase 1

(b)

phase 2

5-Fatigue module FC (Fatigue Cracks)

This module is used to calculate the variation in the rate of fatigue crack propagation (da/dN), and the stress intensity factors in modes I and II (K_I and K_{II}), from the mesh file and the results file (displacements and stresses).

This module has two-routines, the fatigue routine (fatigue crack propagation) and fracture routine (calculation of the stress intensity factors at the tip of the crack).

Two routines:

5.1-Fatigue routine

In order to take into account the phenomena observed during the propagation of a crack, after the application of a constant or variable cyclic loading , we simulated the advance of the crack by freeing the nodes, such that they no longer constrained to remain on the axis of propagation of the crack.

For this , it is necessary to use a node relaxation procedure to make the crack advance.

There are several node relaxation methods. Of these, the method chosen was one based on the change of the boundary conditions by the imposition of a vertical force (vertical displacement); a spring is introduced at each node situated beyond the tip of the crack, in the direction perpendicular to the axis of propagation of the crack, to satisfy the boundary conditions imposed:

Two springs are applied for each mode in the system, corresponding to two degrees of freedom. The spring stiffnesses applied of them are as follows :

+ for a fixed node : the stiffness of the spring is equal to an extreme value (Res= f(cte, Modulus of Elasticity).
+ for a free node : the stiffness of the spring is equal to zero (Res=0).
  where cte is a coefficient which depends on the material used, for example , for steel, cte is equal to a value between 10^3 and 10^8 .

The data required is as follows:
+ number of the node and the element corresponding to the tip of the crack;
+ number of the node and the element along the axis of propagation of the crack (figure 4).

The results:

It is assumed the crack advances in successive steps, each step being equal to the length of an element of the mesh, when the distance travelled by the leading edge of the crack (CTOD: the bottom of the crack has been defined in the program as the zone between the tip of the crack and the preceding node)(figure5) reaches a threshold value which is greater than the non-propagation threshold value corresponding to the threshold value of the stress intensity

factors (ΔK_{th}) ,the node situated at the tip of the crack
is freed ,thus crack propagation takes place,and the rate
of crack propagation has been defined. In the opposite case
where the node is not freed,there is no propagation and
the results show a stoppage of the crack (crack arrest).

The results obtained are the distance travelled by the leading edge of the crack and the rates of propagation of the crack.

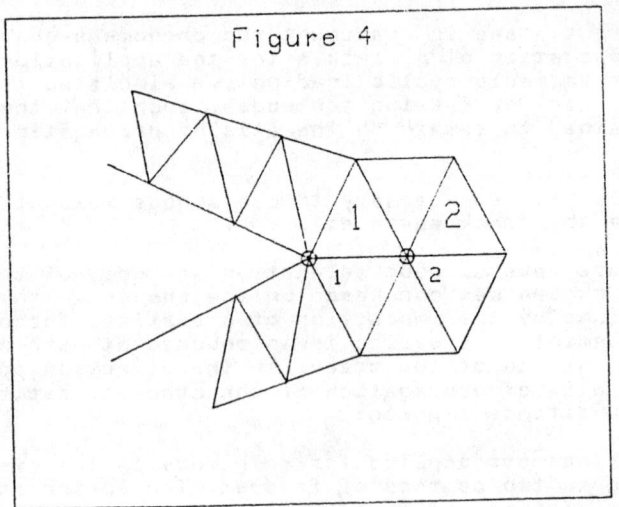

Figure 4

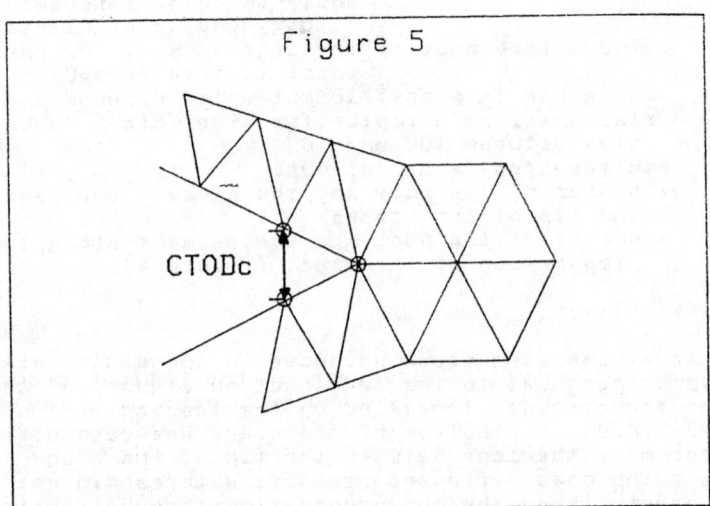

Figure 5

5.2 - Fracture routine

This sub-routine is used to determine the different parameters of the mechanics of the fracture, such as the stress intensity factors in modes I, II, I+II (K_I and K_{II}).
The factors K_I and K_{II} are calculated from the displacement of the node at the lips of the crack.
The data required is as follows:
+ the number of the node corresponding to the first point on the lip of the crack;
+ the number of the node corresponding to the bottom of the crack (the tip);
+ the number of the node corresponding to the 2nd point on the lip of the crack;
+ the angle between the crack and the axis x (figure 6).

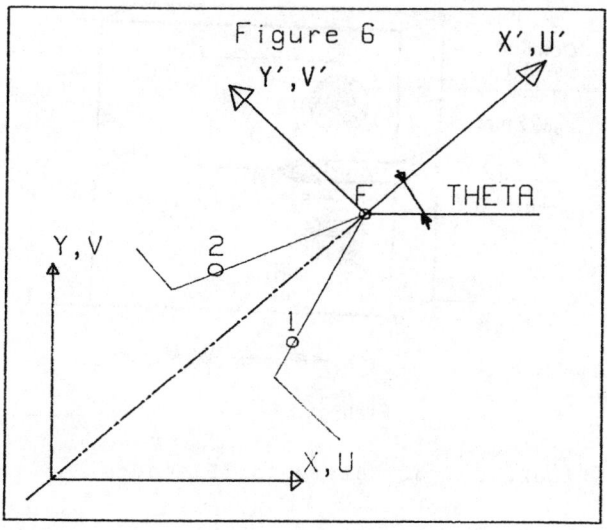

The results:
The results are the stress intensity factors K_I and K_{II}

EXAMPLE OF THE APPLICATION

Numerical simulation of the progress of a crack after the application of a variable loading:
The estimation of the progress of the fatigue cracks in actual structures requires the knowledge of a law of crack propagation which links the rate of propagation of the fatigue crack to the stress intensity factor.
The basic law for simple loadings of constant amplitude is Paris' law (1965); this is no longer valid if the loading is variable. This is due to the "memory" effect in the plastic

zone, at the tip of the crack. Several models have been suggested to take this memory effect into account : the introduction of the dimension of the plastic zone (e.g. Wheeler's model -1972), the notion of the effective stress intensity factor (e.g. Newman's model -1977).
The object of this work was to calculate the changes in the plastic zones and the variation in the rate of propagation of the crack, after the application of the overload, using the method of finite elements, and to compare the result of this calculation with experimental results.

Nature of the simulated tests

The tests were carried out on CT40 compact test specimens (figure 7) in E36 steel.

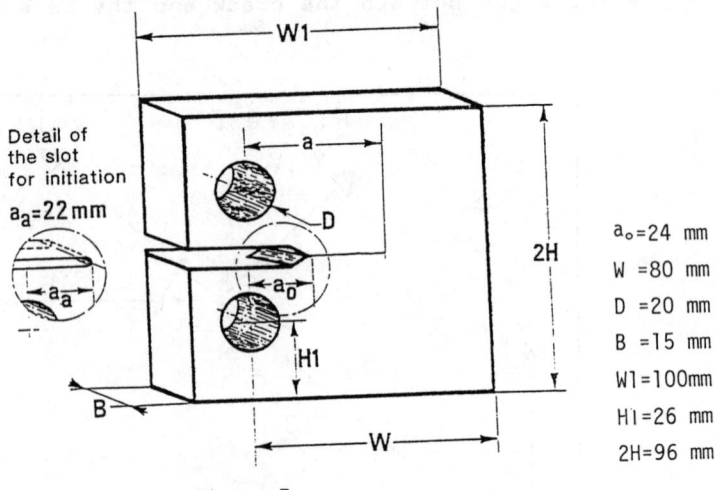

$a_0 = 24$ mm
$W = 80$ mm
$D = 20$ mm
$B = 15$ mm
$W1 = 100$ mm
$H1 = 26$ mm
$2H = 96$ mm

Figure 7.

Mesh (figure 8): 627 elements, 359 nodes.

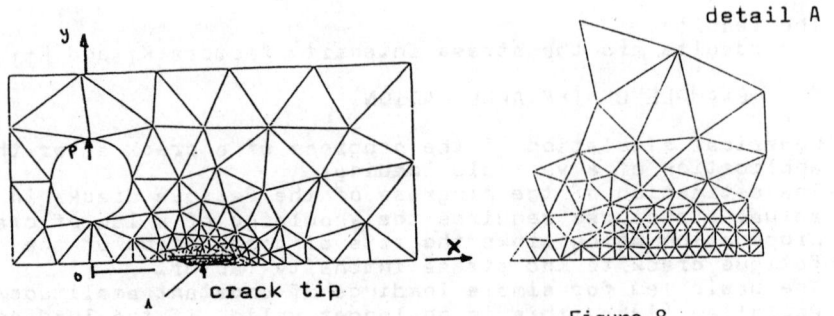

Figure 8.

Characteristics of the material
$E = 20600 \text{ daN/mm}^2$
$\nu = 0.3$
(E36 steel with a tensile yield stress of 350 daN/mm^2, figure 9) ,(monotonic and stabilised cyclic given by Robin et al 1982).

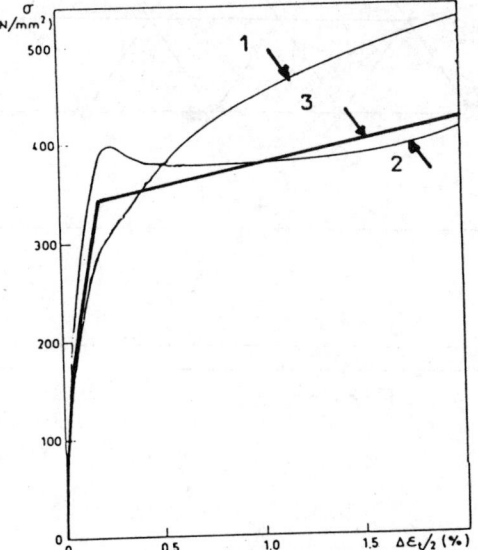

1-Monotonic
2-stabilised cyclic
3-prediction .

Displacement boundary conditions:

The nodes situated along the prolongation of the crack are constrained to move along its axis. The nodes of the crack proper are entirely free.
In order to take into account the phenomena observed in the propagation of a crack after the application of a constant or a variable cyclic loading, the advance of the crack is simulated by the freeing of the nodes.
For this, a node relaxation procedure to advance the crack must be used.
These conditions are achieved simply by associating two springs, aligned respectively along the directions X and Y, with each node situated beyond the tip of the crack (figure 10). The stiffness of these springs is zero where the displacements are free, and very large (10^6 multiplied by modulus of elasticity of the material in the structure) if the opposite is the case.

Loading configuration :

The applied loads are cyclic (figure 11), with the ratio between the maximum and the minimum load equal to 0.1 , and the ratio of the overloads being Rp=1.9 and 2.2 .
The maximum load corresponds to the maximum stress intensity factor $K_{max} = 18 \text{ MPa.(m)}^{1/2}$.

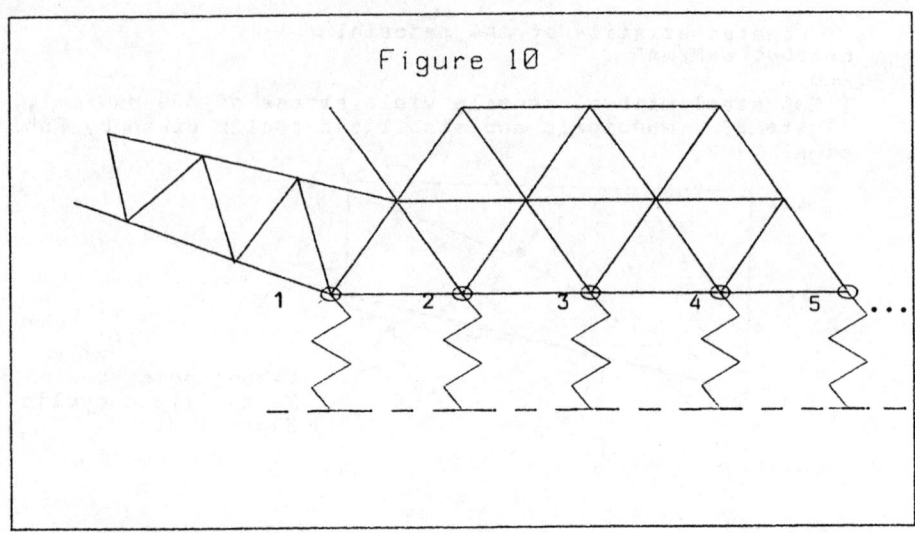

Figure 10

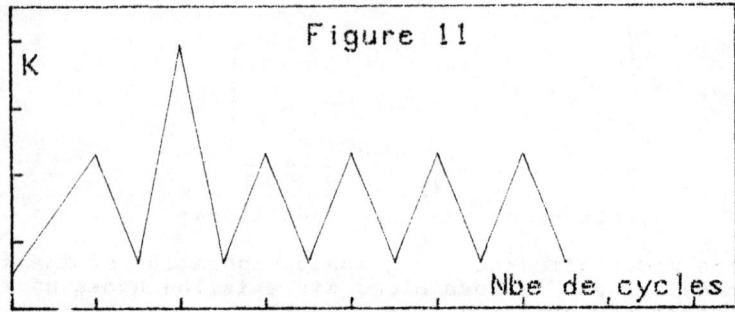

Figure 11

Propagation criteria:

When the distance travelled at the bottom of the crack CTOD (the bottom of the crack has been defined in the program as the zone between the tip of the crack and the preceding node)exceeds a non-propagation threshold value,the node situated at the tip of the crack is freed.The crack thus advances in successive steps,each step being equal to the distance between two succeeding nodes (figure 1o).With such a criterion the rate of crack propagation can be defined.
The non-propagation vertical displacement threshold for the first node situated in front of the tip of the crack is calculated from the formula proposed by Bia (1985):

$$CTOD_{th} = 0.425 \cdot (\Delta K_{th}/E \cdot R_e) \quad (1)$$

where R_e is the tensile yield stress,
ΔK_{th} is the amplitude of the intensity factor at the non-propagation threshold.
If propagation takes place,the rate is controlled by a law of the type:

$$da/dN = A \cdot (CTODc)^B \qquad (2)$$

with $A = C \cdot (E \cdot Re/0.73)^2$
and $B = 2m$

where C and m are constants which depend on the nature of the material, and CTODc is the vertical displacement given by the FE calculation.

Results:

In particular, the calculations gave:
+ the amplitude of the opening of the crack;
+ the contour of the plastic zone;
+ the variation of the rate of crack propagation.

The opening of the crack, represented by the vertical displacement of the nodes situated in the plane of symmetry and in front of the tip of the crack.
In table 1, we compare the results found by analytical methods (Rice's and Dugdale's models) with the result found by the method of finite elements. (In order to determine this point by the FE calculation, it will have been necessary to carry out intermediate calculation steps between the minimum and maximum loads).
The close agreement between the calculations and the two analytical models proposed (Rice's and Dugdale's models) will be noted.

	During overload	After overload		
a (mm)	30,57	30,63	....	31,06
CTOD-Rice μm	12,95	2,08		
CTOD-Dugdale μm	17,95	4,9		
CTOD-EF μm	12,08	4,65	....	5,79

Table no.1 : Calculation of the distance travelled by the crack during and after overload (Rpeak = 1,9)

The variation in the rate of crack propagation by the method of finite elements, together with the experimental results (Robin et al -1982), are shown on figures 13a and b.
It will be noted that the method of finite elements gives close agreement with the experimental results, and forecasts the stoppage (or slowing-down)/ crack arrest or delayed re-

tardation/in the progress of a crack, under experimental conditions where steppage/crack arrest/(slowing-down/delayed retardation/)actually occurs.
The experimental parameters-the minimum rate of crack growth after overload $(da/dN|min)$, the length of crack at which the minimum rate appeared $(a_d|min)$, the length of crack affected by the overload (a_d^*), and their calculation equivalents-are given in table 2.

		$da/dN|min$ (mm/cycle)	$a_d|min$ (mm)	a_d^* (mm)
Rpeak=1,9	Experimental measurements	$4,5 \cdot 10^{-7}$	0,28	0,8
	EF calculations	$0,33 \cdot 10^{-7}$	0,24	0,72
Rpeak=2,2	Experimental measurements	Crack stopped	0,4	---
	EF calculations	Crack stopped	0,42	---

Table no.2 : Comparison of the experimental results with the calculations by the method of finite elements.

Conclusion

The simulation by finite elements is revealed as a useful tool for the simulation of the propagation of a crack which is subjected to complex loading conditions (constant or variable). Close agreement has been obtained between simulation and practice, by combining the numerical procedure for the progression of the crack and the physical phenomena controlling it, notably the opening of the crack. This method constitutes an improvement over the procedure of the advance of the crack, node by node.

REFERENCES

Bia A. (1985), CETIM   Internal Report
Frederic C.O., Wong Y.C., and Edge F.W. (1970), Int.J.Num.Mech. Vol.2, 133-144
Kaiser S., and Carlson A.J. (1983), ASTM-STP 803, II, 58-79
Kanninen M.F. (1973), Int.J.Fract, Vol.9, No1, 83-92
Lawrence L.D. (1979), Comp.& Struct.Vol.10, 561-575
Marcal P.V., and King I.P. (1967), Int.J.Mech!Sc.Vol.9, 143-155
Newman J.C.Jr, (1977), ASTM-STP 637, 56-80
Paris P.C., and Sih G. (1965) ASTM-STP 381, 30-83
Robin C., Chehimi C., and Pluvinage G. (1982), ICSMA6, Vol.2, Melbourne August, 919-926
Robin C., Chehimi C., Louah M., and G.pluvinage G.(1982), Proc. 4th E.C.F., 488-494
Thomas T.J., Nair S., and Garg V.K. (1983), Comp.& Struct., Vol. 16, No.5, 669-675
Wheeler O.E. (1972) ASM.J.Basic.Eng.94., 181-186
Zienkiewics O.C et al (1969) Int.J.N.Me.Eng.Vol.1, 75-100

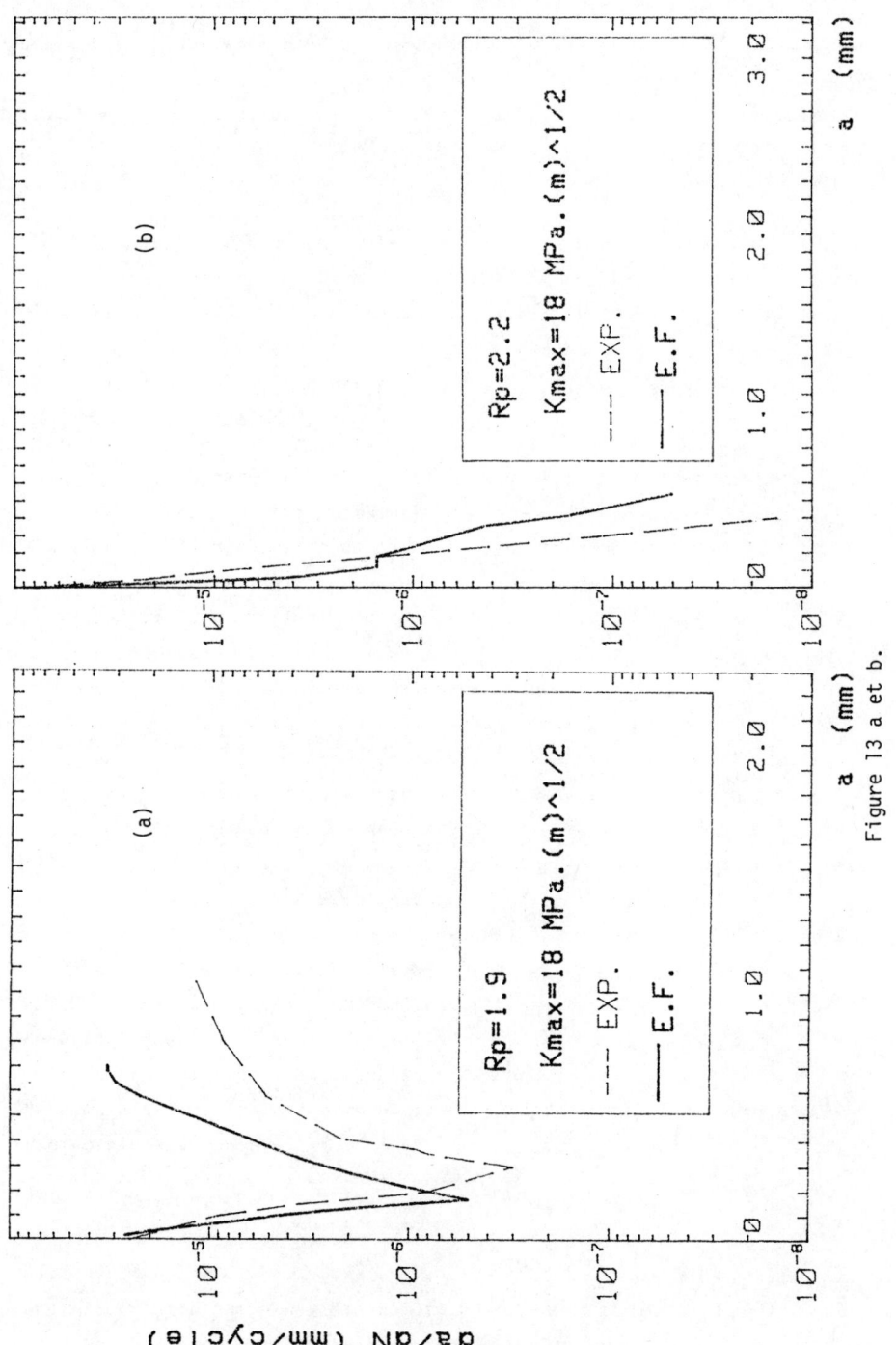

Figure 13 a et b.

THE BOUNDARY ELEMENT METHOD AS A CAD TOOL

C. A. Brebbia

Computational Mechanics Institute, Southampton, UK

BEM and the CAD Process

Over the last few years the engineering industry - in particular mechanical engineering - has accelerated the implementation of Computer Aided Design codes. This tendency which is dictated by considerations of efficiency and productivity will continue within the forseeable future and is hence important for analysts to understand how analytical codes will integrate with CAD systems.

The design of mechanical components using computers begins with the description of the particular piece or structure under consideration applying a numerical modelling system. Once this is done the piece can be displayed and its different views, cross sections, juxtapositions, etc. can be studied on a graphics terminal. The next stage is the analysis of the component under the type of force or temperature state, under consideration. This, up to now, has been done almost exclusively using finite element codes but more recently engineering users have become aware of the potentialities of boundary elements. Whatever technique one uses, a preprocessing code is needed at this stage to link the geometric module to the analysis package. These codes produce the input data required to run the analysis module, such as element definition, geometry, loading, boundary conditions, etc. Once the analysis has been completed the results need to be presented in a form which is easy to interpret. This can be done using a post-processing graphics package and special emphasis is nowadays laid on the use of interactive colour graphics.

As the first analysis is seldom the final one, the next stage is an interactive process, during which the designer makes certain modifications to the computer model until the desired stresses or temperature profiles are obtained. The final data file can be plotted using a drafting system to obtain the blueprints required in the manufacturing process, or it can be used to operate a CM or robotics facility. The process is diagrammatically shown in figure 1.

Although the design process has in this manner been computerized, it is still highly dependent on the skill of the designer himself who is responsible for obtaining an optimum solution. As the process is dependent on the individual it is desirable to have as interactive a system as possible. CAD may however change in the foreseeable future if the designer's skill can be incorporated in the computer system itself. In this regard recent developments in Artificial Intelligence may influence the design process. CAD codes could for instance be able to let the user know if the answers are reasonable, the results meaningful, if more iterations are required, how to proceed in order to find an optimum solution, etc. To achieve this the program ought to be able to learn from experience and artificial intelligence tools such as data base management and optimization techniques will need to be implemented in present data codes. Although this is a promising field of research, no major codes utilize any of these concepts at present.

Of more important direct relevance to the designer is the development of analysis packages better suited to the type of interactive process involved in CAD. Finite element codes were developed many years ago when there were no requirements for CAD and they tend to be cumbersome to use. Boundary elements instead, offer important advantages over finite elements including being much more efficient solution in terms of man-time. Not only can meshes be easily generated but what is perhaps more important, design changes are simple to include as generally they do not necessitate a complete remeshing. Other advantages of boundary elements is that the technique gives more accurate results in regions of stress or flux concentration and the method is ideally suited to treat problems with infinite domains.

In this paper the advantages of BEM as a CAD tool are described in detail through a series of engineering applications. The objective is to demonstrate the potentialities of the technique in engineering practice.

Awareness of the advantages of BEM should be coupled with an understanding of the technique. In this regard it is important to point out that the simplicity of BEM for the user has its foundations on a more complex mathematical approach. This complexity which if properly resolved contributes to give more accurate results in well written BE codes can easily result in inaccurate results in poorly written codes. The BEM is generally more susceptible to errors when not using the appropriate numerical techniques and users are well advised to check the accuracy and convergence of the results produced by new BE codes. This advice - which is frequently ignored - applies of course to FE computer codes as well. It is remarkable to see how few quality and reliability checks of engineering software are being carried out.

This paper is addressed to BEM users or potential users and will try to answer the questions which they more frequently raise regarding the quality of BEM solutions and how they compare with FEM. In particular the paper discusses the following topics:

* Are BEM solutions accurate and convergent?
  (or should I bother about it?)
* How well BEM compares with FEM?
  (or should I tell the boss about it?)
* Can BEM deal with infinite problems?
  (or how infinite is a finite element?)
* Has BEM serious advantages for the designer?
  (or should I use it now or wait?)
* What the BEASY code can do
  (or should I believe the marketing guys?)

Are BEM Solutions Accurate and Convergent?

The most attractive feature of BEM for the designer is its ability to solve problems using only the boundary data. This implies that the method requires less data than FEM and consequently reduced effort to set up the problem. In spite of this it is of primary importance first to assess the performance of the analytical technique from its accuracy and convergence viewpoint. Accurate solutions are needed for problems such as those for which stress concentrations may occur. Convergence is a basic requirement as otherwise the code could hide some fundamental mathematical errors which will surface sooner or later in the results.

In general accuracy of the solution versus degrees of freedom and time required to prepare and run a particular case gives the designer a yardstick of how to use the code. A program which requires highly refined meshes in order to obtain the desired accuracy is generally unattractive versus a code which only requires the minimum information.

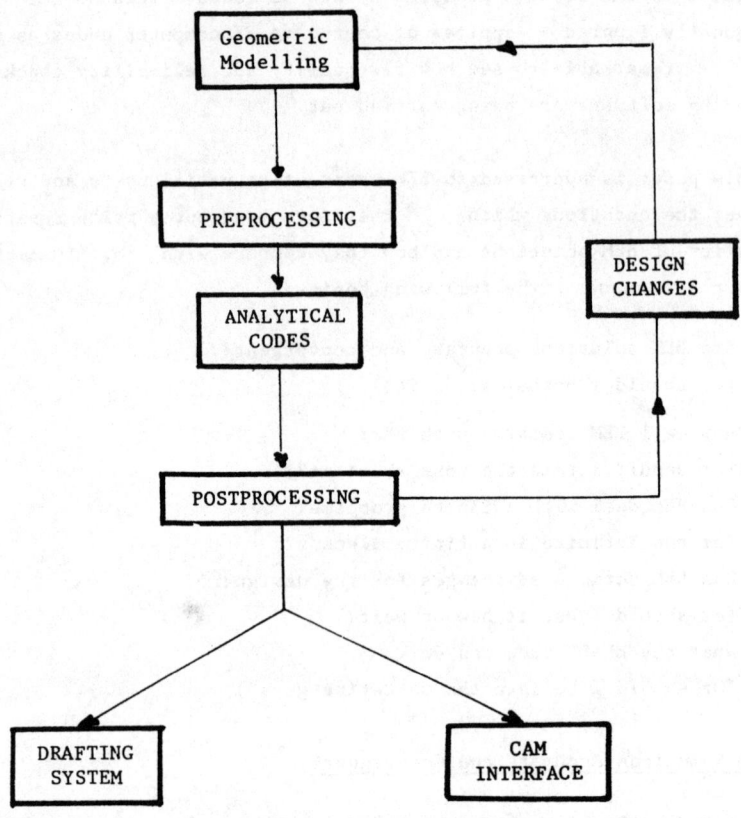

Figure 1 The CAD Process

One can illustrate these points by looking at some simple examples which give an idea of the numerical performance of boundary elements. Results will usually be compared against finite element programs and in particular against a well known British code (all have FEM results). The Boundary Element code BEASY * has been written by Computational Mechanics Institute at Southampton, UK. The set of results for the cantilever beam were obtained using the same VAX 11/730 computer which is part of the CAD training facilities at the Institute. The versions used here were implemented by the companies marketing the codes.

The first example analysed was a cantilever beam (in plane stress), clamped at one end and with a uniformly distributed load at the other end as shown in figure 2. This case is of interest in engineering as it combines significant bending deformations with some due to shear, for the case of height to length ratio equal 1 to 4. The beam theory values for tip vertical deflection (at A) and two representative stresses (at B and C) are also shown in figure 2. Numerical results should approximate these values within a certain error. Due to the presence of shear deformation (neglected in the classical beam theory) one would expect the actual deflections to be slightly larger than those predicted by the beam theory. Direct stresses will be slightly smaller for σ_B and larger for σ_C due to their non-linear distribution across the thickness.

The beam was discretized as shown in figure 3 using 8 node quadrilateral model for the finite element codes (PAFEC and LUSAS) and discontinuous quadratic elements for boundary elements (BEASY program). Results for

---

BEASY. Apparently the only complete and fully available Boundary Element Code, which can solve potential as well as stress analysis problems. It was developed by Computational Mechanics in the UK under the supervision of Dr. Brebbia (Version 2.19).

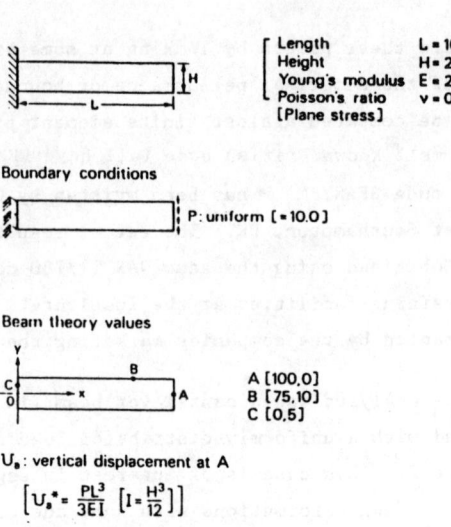

Length L = 100
Height H = 20
Young's modulus E = 2×10⁴
Poisson's ratio ν = 0.3
[Plane stress]

Boundary conditions

P: uniform [= 10.0]

Beam theory values

A [100,0]
B [75,10]
C [0,5]

U_a: vertical displacement at A

$$\left[U_a^* = \frac{PL^3}{3EI}\left[I = \frac{H^3}{12}\right]\right]$$

σ_b: direct stress at B

$$\left[\sigma_b^* = \frac{1}{4} \times PL/[I/H/2]\right]$$

σ_c: normal stress at C

$$\left[\sigma_c^* = \frac{1}{2} \cdot \frac{PL}{H}\right]$$

Fig. 2 Example problem 1

	BEM MODEL (1 zone)	BEM MODEL (2 zone)	BEM MODEL (4 zone)	FEM MODEL
2×1	▭	▭▭	—	▭▭
4×2	▭	▭▭	▭▭▭▭	▭▭▭▭
8×4	▭	▭▭	▭▭▭▭	▭▭▭▭▭▭▭▭
16×8	▭	▭▭	▭▭▭▭	—

1. BEM (BEASY)   ⌐⌐ discontinuous quadratic element
2. FEM   ▱   8 nodes quadrilateral element

Fig. 3 Model discretization for cantilever beam

		d.o.f.	Ua/Ua*	σb/σb*	σc/σc*	CPU time
B E A S Y	1 zone	2x1	36	1.195	1.221	1.715	37"
		4x2	72	1.013	0.985	1.335	1'40"
		8x4	144	1.013	0.986	1.383	5'29"
		16x8	288	1.021	0.994	1.343	34'15"
	2 zone	2x1	42	1.367	1.297	1.890	44"
		4x2	84	1.041	1.000	1.370	2'00"
		8x4	168	1.022	0.996	1.394	6'14"
		16x8	336	1.024	0.997	1.347	39'01"
	4 zone	4x2	108	1.091	1.051	1.447	2'54"
		8x4	216	1.029	1.001	1.408	14'12"
		16x8	432	1.025	0.998	1.349	53'12"
F E M		2x1	26	0.969	0.996	1.403	6'56"
		4x2	74	1.016	0.991	1.520	7'27"
		8x4	242	1.024	1.000	1.427	12'20"

CPU time : on DEC VAX 11/730

Table 1 Results for cantilever beam bending

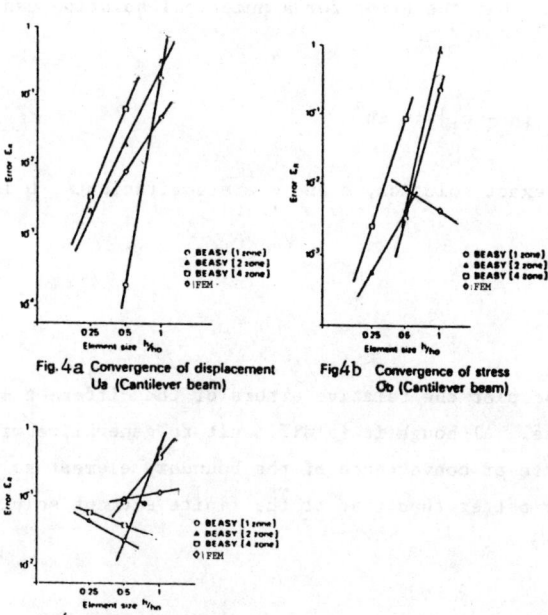

Fig. 4a Convergence of displacement Ua (Cantilever beam)

Fig. 4b Convergence of stress σb (Cantilever beam)

Fig. 4c Convergence of stress σc (Cantilever beam)

the three codes are shown in table 1, which gives the ratio between the solutions obtained using different meshes and the beam theory results (indicated by an asterisk). Boundary element results are also presented for different subdivisions into regions to investigate if the subdivisions have any bearing on the accuracy of the solutions. In general this was not the case although the subdivisions affect the computer time, in this case unfavourably as the number of regions is small. If they increase computer time should be reduced but in the present application the bookkeeping needed in the code outweighs the advantages of using sparse matrices.

In order to study the convergence of the solution it was decided to plot a measure of the <u>relative error.</u> For a particular value, say a u displacement, this error was defined as,

$$\varepsilon_i^R = |u_{i+1} - u_i| \qquad (1)$$

where u_i is the solution for mesh 'i' and u_{i+1} for the next and improved mesh, i+1 as given in table 2.

Notice also that the error for a numerical solution can be generally expressed as

$$\varepsilon_i = |u - u_i| = ch^\alpha \qquad (2)$$

where u is the exact solution, c and α are coefficients h is the element length.

Figures 4a to 4c plot the relative errors of the different solutions versus the element size. Although it is difficult to generalize with only one example, the rate of convergence of the boundary element solution appears to be generally better than that of the finite element solution obtained with the FE program.

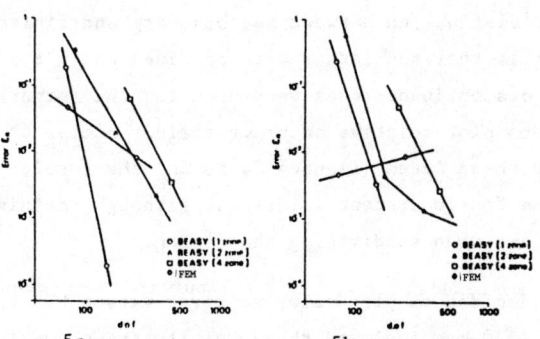

Fig. 5a Convergence of U_a vs d.o.f. (Cantilever beam)

Fig. 5b Convergence of σ_b vs d.o.f. (Cantilever beam)

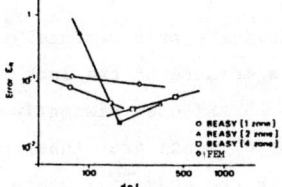

Fig. 5c Convergence of σ_c vs d.o.f. (Cantilever beam)

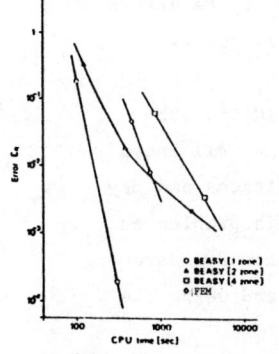

Fig. 6a Convergence of U_a vs CPU time (Cantilever beam)

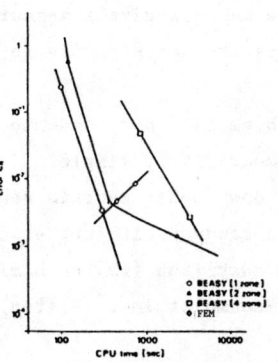

Fig. 6b Convergence of σ_b vs CPU time (Cantilever beam)

		DIVISION	$\Delta\sigma_b/\sigma_b^*$
B E A S Y	1 zone	4x2	0.105
		8x4	0.020
	2 zone	4x2	0.109
		8x4	0.022
	4 zone	4x2	0.111
		8x4	0.023
FEM		4x2	0.116
		8x4	0.040

$\Delta\sigma_b$: Discontinuity of σ_b

Table 2 Discontinuity of stress (Cantilever beam)

Fig. 6c Convergence of σ_c vs CPU time (Cantilever beam)

An important distinction between the boundary and finite element results however is that the latter were obtained using continuous elements while discontinuous ones were used for the former. Consequently it was decided to plot relative accuracy against number of degrees of freedom and for these cases (figures 5a to 5c) the results also compare well against the finite element solutions, although certain accuracy appears to be lost when subdividing the region.

It was also decided to plot relative error versus CPU time (using the same VAX 11/730 machine) and these results are shown in figures 6a to 6c.

As a final test the stress discontinuity at B was analysed. This discontinuity-called here $\Delta\sigma_B$-gives a measure of the lack of equilibrium present in the approximate solution. BEASY uses discontinuous elements, so that σ_B values from two different elements are inherently different. (They are as a matter of fact, used in the code as a measure of the convergence of the solution). In the case of finite elements, nodal stresses produced by one or other element joining at B will also be different. The results shown in table 2 give a measure of the errors in equilibrium at B. The results are satisfactory in all cases.

The problem of a cantilever beam is a bit academic and the subdivisions studied above were comparatively simple. It is well known that certain BE codes can break down under certain conditions and may not give convergent results. To study a bit further this problem an industrial example was run by Wanderlingh [13] at Hamilton Standard with the help of Computational Mechanics Inc. (Boston) and using the BEASY code.

The structure used was the gear tooth shown in figure 7 where the starting mesh of 1060 constant strain finite elements is also shown. A point load was applied to the tooth pitch diameter and the boundary conditions were zero displacements all around the base. The first boundary element model consisted of only 41 quadratic elements as shown in figure 9. The boundary element method peak is 5% higher than that of

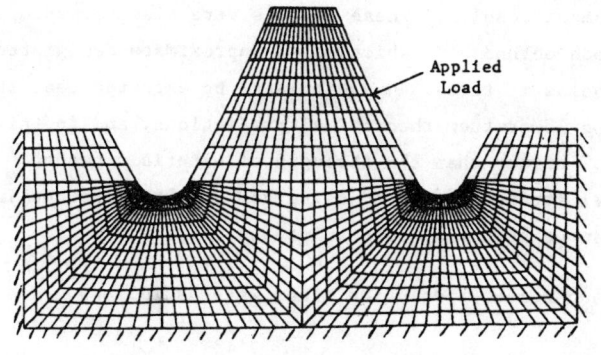

Fixed Boundary

FIGURE 7

FINITE ELEMENT MODEL OF INTERNAL SPUR GEAR TOOTH

Element Identification Numbers are Shown

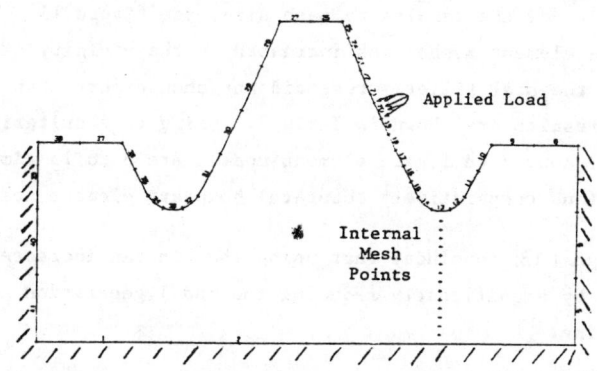

Fixed Boundary

FIGURE 8

BOUNDARY ELEMENT MODEL OF INTERNAL SPUR GEAR TOOTH

the finite element results. These results were also compared with a Modified Heywood method which is an approximate design technique based on photoelastic tests. Hence it is to be expected that the latter results will be lower than theoretical predictions, and in this case the BE solution is 15% higher than the strength of materials method. Comparison of the internal stresses obtained using FEM and BEM is shown in figure 10 where it can be seen that both solutions correlate well.

Modelling and computer solution times were as follows.

	Model generation time (min)	Lines of input	CPU (sec) Solution time
FEM	60	130	18
BEM	20	33	19

(The above analyses were carried out on an IBM 3084 computer).

It is important to point out that the average modelling time using BEASY preprocessor was approximately three times faster than the FE modelling time.

Finally, several models of the gear tooth were constructed to determine the sensitivity of the results to mesh size, see figure 11. The density of the finite element meshes was increased in the vicinity of the tooth fillet until the peak fillet stress did not change more than 5%. Convergence results are shown in Table 3. Using this criteria the CPU times for the converged finite element models are 3 to 15 times higher than the initial (and comparatively accurate) boundary element solution.

Wanderlingh [13] concluded that using the BEM can increase engineering productivity by significantly reducing the model generation and data production time.

How Well BEM Compares with FEM?

The finite element method is now a well established technique for stress analysis, having started approximately 25 years ago and with a series of major computer systems, the oldest of which was first released in 1967. Originally the method was used on its own, i.e. independent of any

FIGURE 9

TOOTH FILLET SURFACE STRESS

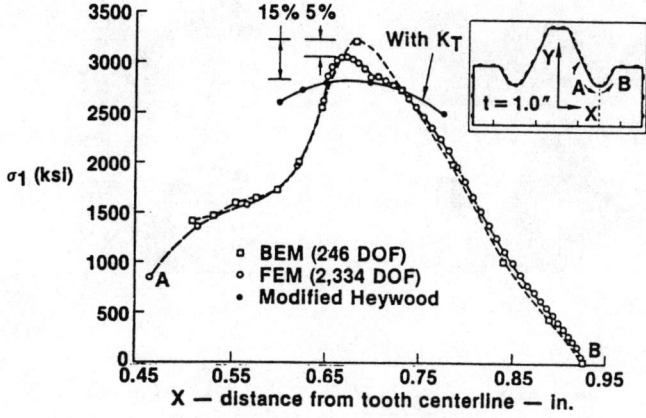

FIGURE 10

INTERNAL TOOTH STRESS

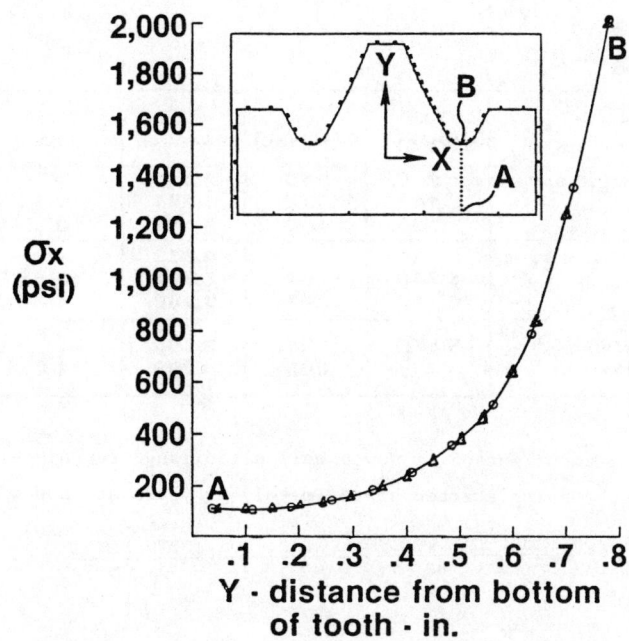

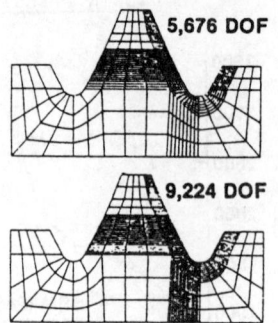

FIGURE 11

FINITE ELEMENT TOOTH MODELS

Code	Size (DOF)	CPU (sec)	σ max (psi)	Δ %
BEM-Beasy	246	19	3,210	-
	444	112	3,184	0.8
	528	164	3,193	0.3
Fem-in-house	2,334	18	3,053	-
(Standard-	5,566	67	3,374	10.5
Hamilton)	6,112	61	3,232	4.2
FEM-MSC/	5,676	164	3,260	-
Nastran	9,224	280	3,299	1.2

Table 3

MAXIMUM STRESS CONVERGENCE FOR TOOTH MODEL

other computer packages, but more recently it is seen as part of the
CAD/CAM process, with the current emphasis on integration of the codes
with other CAD programs. This tendency, which applies to other analytical
techniques including boundary elements, is now accelerating and more
analysis codes are in the process of being interfaced to existing CAD
systems.

In spite of FEM being well established by now and of having had an
overwhelming success in the engineering environment, the technique has
some serious drawbacks. FEM meshes require discretization of the entire
volume of a structure, refined meshes are required in regions of stress
concentration and meshing errors are difficult to detect, particularly
in three-dimensional models.

FE models are nearly always based on the principle of minimum
potential energy and they work with the displacements as the problem
unknowns. This produces the classical behaviour of stiffness solutions,
i.e. their overall rigidity versus the correct solution. Because of
this, results for displacement solutions such as finite elements are said
to approach the correct solution 'from below'. This property can be used
when trying to assess the rate of convergence of approximate finite
element results, but to do so will require obtaining several numerical
solutions for varying degrees of mesh refinements. This approach is
unfortunately useless in engineering practice where the mesh is most
frequently determined by the geometry of the problem. Hence it is always
difficult to assess the accuracy of the solution and the FEM approximation
is always liable to give results corresponding to a 'stiffer' problem,
which unfortunately introduces a non-conservative error. Few FE codes
have any checks regarding their accuracy, although it should be
comparatively simple to test for equilibrium of the solution, for instance.
Numerical results however tend to indicate that many FE solutions satisfy
equilibrium only in a very approximate manner. This is more clearly
seen when trying to interpret boundary forces or reactions obtained using
finite element codes.

The boundary element method is a more recent technique than FEM.
Although the basic theory of integral equations was well known since the
beginning of this century its numerical implementation was dependent not
only on the advent of modern computers but also on the development of

new numerical techniques. Many of these techniques originated with finite elements but were further developed before being implemented in boundary element codes. Typical examples are numerical integration methods, solutions of sparse matrices and others.

The main advantage of boundary elements versus classical domain techniques such as finite elements or finite differences is that it only requires the discretization of the boundary. This implies that it is easier to prepare its data and consequently the codes are simple to interface with CAD systems. The price to pay for this simplicity of use is the inherent complexity of the mathematics behind the method. This in part explains the difficulties of extending boundary elements to deal with non-linear problems, in which case the method requires internal points definition, losing some of its attractiveness.

Boundary elements are better conditioned to solving problems extending to infinity for which finite elements are obviously unsuited. The technique is then ideal for applications in geomechanics, foundation engineering, some hydraulics and water wave studies, acoustics and other field problems. Another important advantage of boundary elements versus finite elements is that the method is based on mixed rather than potential only principles. This implies that solutions obtained with the method are not a lower - i.e. ' stiff' - bound but tend to lie in between the lower and upper bounds. Numerical results obtained using BEM are then inherently more accurate. In addition the mixed character of the method implies that both displacements and surface stresses are considered as variables when solving stress problems. Hence both are given with the same degree of accuracy while in finite elements it is necessary to find numerically the derivatives of the displacement functions in order to compute the stresses, an operation which inherently produces a loss of accuracy in the results.

Table 4 summarizes some of the advantages and disadvantages of both techniques. One of the aims of this paper is to assess qualitatively their numerical performance. Towards this end results obtained using different BEM/FEM programs are compared.

	FEM	BEM
Advantages	* Simple Mathematical Theory * Easy to extend to non-linear problems	* Boundary Only requires discretization * Accurate results in regions of stress concentration * Better suited for infinite regions * Simple data preparation and consequently interface easily with CAD systems
Disadvantages	* Entire volume needs to be discretized * Refined meshes in stress concentration regions * Not well suited for infinite regions * Complex data preparation and checking	* Complex mathematical formulations * Difficult to extend to non-linear problems

Table 4 Advantages and Disadvantages of FEM/BEM Methods

Sometimes the FE solutions are unreliable for engineering applications. A particular case occurs when analysing nearly incompressible materials. Consider the cylinder under constant pressure shown in figure 12. The structure can be considered to be under plane strain and the Poisson ratio was taken to be 0.49999. (BEASY can in principle analyse the case $\nu = 0.5$ but this may be thermodynamically undefined). Results using BEASY show a reasonable behaviour although the solution for σ_θ deteriorates when $\nu = 0.49999$. (see Table 5). Two FEM codes were tried to analyse this case. One of the FE packages studied gave considerable error for the case $\nu = 0.49$.

Many mechanical engineering pieces present sharp corners or regions of stress concentration and it is important to have reliable results for these cases. Figure 13 shows a test specimen with a groove subjected to a force acting in the longitudinal direction. Because of the geometry

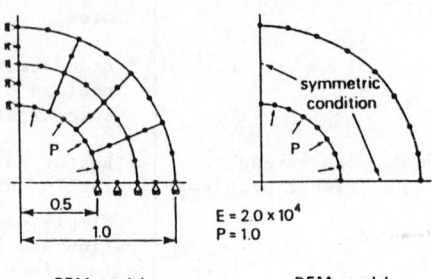
Fig.12 Nearly-incompressibility test

Stress σ_θ

ν	analytical	BEASY	FEM 1	FEM 2
0.3	2/3 (0.6̇)	0.66666	0.654	0.666349
0.4	↑	0.66667	0.675	0.666379
0.49	↑	0.66674	1.04	0.666379
0.499	↑	0.66750	4.30	0.666379
0.4999	↑	0.67516	17.8	0.666349*
0.49999	↑	0.70278	26.4	0.666349*

Displacement Ur

ν	analytical	BEASY	FEM 1	FEM 2
0.3	3.0333×10⁻⁵	3.0333	3.0316	3.03368
0.4	2.8000	2.8000	2.7967	2.80037
0.49	2.5333	2.5333	2.5014	2.53340
0.499	2.5033	2.5065	2.2179	2.50373
0.4999	2.5003	2.5322	1.0957	2.50073*
0.49999	2.50003	2.8605	1.8288	2.50043*

* = with warning

Table 5 Results of nearly-incompressibility test

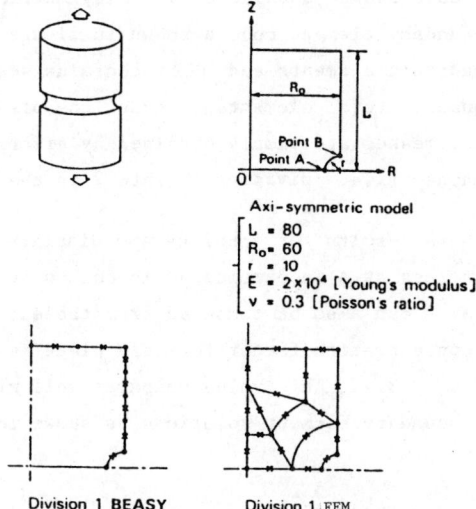

Fig. 13 Specimen with a groove (Axi-symmetric case)

		σz at A	Uz at B	CPU time
B E A S Y	1	4.02	1.63 ×10⁻³	1'12"
	2	3.51	1.64	3'09"
	4	3.36	1.65	10'05"
	8	3.35	1.65	50'32"
F E M	1	2.72	1.50	11'26"
	2	3.11	1.62	14'24"
	4	3.38	1.65	27'15"

Table 6 Results for the specimen with a groove (Axi-symmetric case)

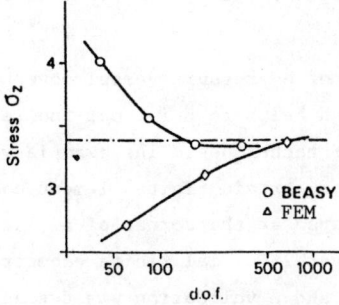

Fig. 14 Stress value at A vs d.o.f. (Axi-symmetric case)

and loading this case can be considered as axisymmetric and solved using
the finite and boundary element models shown in figure 13. BEASY uses
discontinuous quadratic elements and FEM contains second order
triangular and quadrilateral elements to model the piece. Figure 13 shows
mesh 1. The other meshes are simply obtained by discretizing each element
in a monotonic manner (i.e. dividing it into 2 in the case of BEM).

Figure 14 shows results for stresses and displacements at two typical
points. The σ_z stress at A in particular is due to stress concentration.
The value of σ_z at A can also be computed from tables; if so, one finds
that the stress concentration factor for this piece is $K_\alpha \cong 2.4$, which
gives a value of $\sigma_z \cong 3.4$. This value compares well with those obtained
using finite and boundary element solutions as shown in figure 14, and
table 6.

From these results it is possible to plot convergence rates for σ_z
versus element size (figure 15), degrees of freedom (figure 16) or
versus CPU time (figure 17). It is interesting to point out that the
rate of convergence of the finite element results is faster for displacements than for stresses as is to be expected.

Stresses were also compared along the centre line of the specimen
to check their distribution and to see how they equilibriate the applied
forces. Figure 18 demonstrates that although the σ_z stresses behave
reasonably well for both techniques a smoother distribution of the σ_r
values is obtained using BEASY. Finally the surface stresses along the
groove were investigated, one of which (the normal should be zero and
the other decrease from a value of approximately 3.4 to a small value
at the corner (figure 19). For this case the boundary element results
seem to behave better, especially for σ_n tractions which were practically
zero all along the groove.

The following analysis of a pressure vessel cover (figure 20) is an
interesting application which helps to point out the relative advantages and
disadvantages of FEM and BEM techniques. The example was originated by
Floyd [14] as a comparison of certain finite element codes and concerns
the determination of the shapes at the corner of a flat-bed thick-
walled pressure vessel. Several initial finite element analyses gave
different numerical results and a validation was deemed necessary using

143

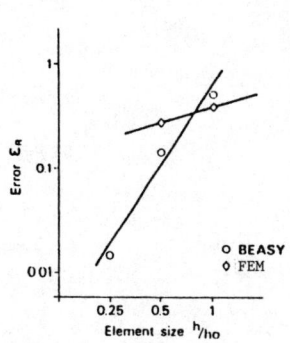

Fig.15a Convergence of σ_z at A

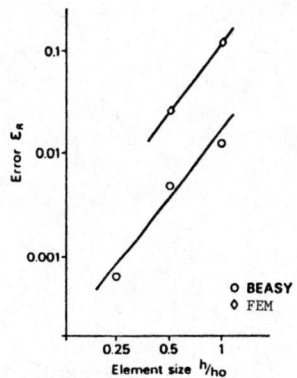

Fig.15b Convergence of u_z at B

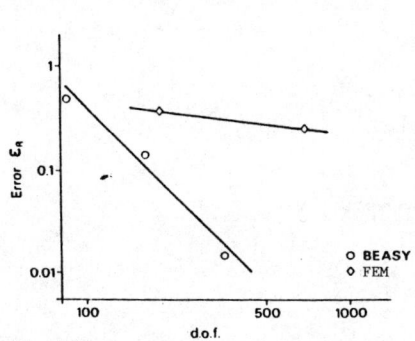

Fig.16a Convergence of σ_z at A vs d.o.f.

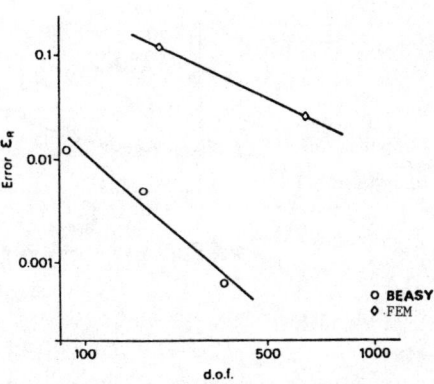

Fig.16b Convergence of u_z at B vs d.o.f.

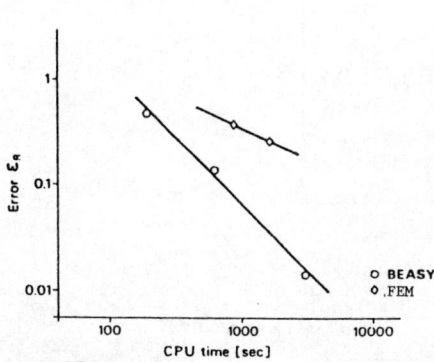

Fig.17a Convergence of σ_z at A vs CPU time

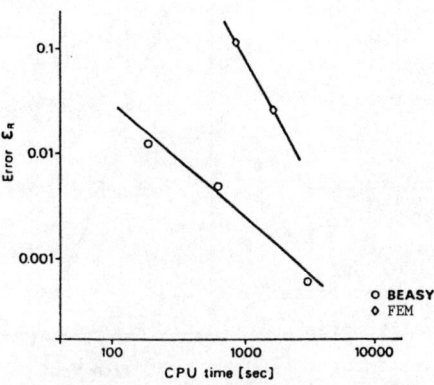

Fig.17b Convergence of u_z at B vs CPU time

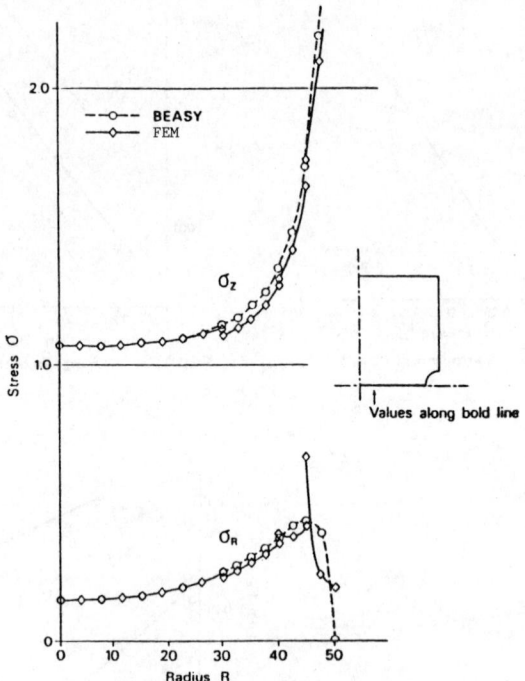

Fig. 18 Stress at internal points (for the fine mesh)

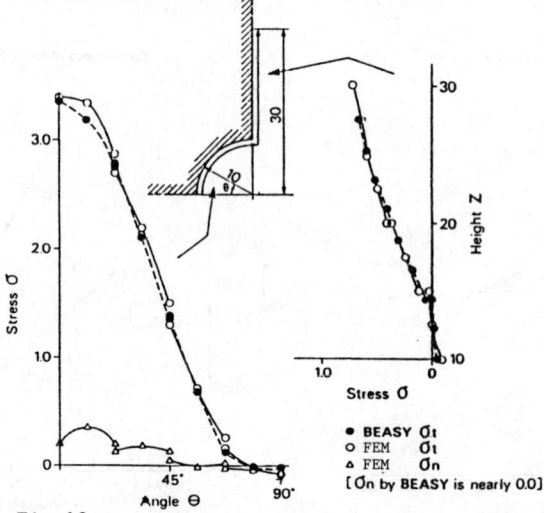

Fig. 19 Surface stress values

an experimental photoelastic model and then the BEASY boundary element
code. The photoelastic model has a Poisson's ratio of $\nu = 0.5$ and was
built out of epoxy resin. The area of interest is the corner fillet
(figure 20) for which the peak surface stress and the corner are
required together with the variation of stresses through the cylinder
wall at the location of the peak stress. The line for which results
were obtained for the photoelastic model is also shown in figure 20,
this line passes through the point of maximum surface stress around the
corner. Floyd reported finite element results obtained using several
codes, for which a Poisson's ratio of 0.4 was used to avoid the errors
associated with large Poisson's ratio. For values over 0.4 he found
overestimation and instability of the solution. One of the set of
finite element results shown by Floyd was obtained using the ADINA
model and the principal stresses plotted along the line defined above.
Local irregularities and inaccurate results were reported for all
finite element results including those with ADINA. By comparison
boundary element results obtained using the BEASY code were in close
agreement with the experimental results. Unfortunately Floyd did not
specify the type of mesh used in the analysis and this gave origin to
some confusion. More recently Bathe [15] attempted to prove that this
ADINA finite element code gave accurate results if the correct mesh
was used and reported close agreement with Floyd's experimental and BEASY
results for a refined finite element mesh. He analysed the pressure
vessel using 69 six noded elements (approximately 500 degrees of freedom),
491 (approximately 2500 degrees of freedom) and 240 elements (around 1500
degrees of freedom). The first discretization gave poor results and
confirmed that the mesh was too coarse. The second case represents
a grossly refined mesh with elements which are badly conditioned because
of their dimensions. The last analysis is the most interesting case
because the mesh is refined in the fillet region and funnels out nicely
towards the regions where stress vary slowly. This mesh gives the best
stress plots of the three meshes and results were then compared against
BEASY results, although the original boundary element results were
obtained for a different Poisson's ratio i.e. $\nu = 0.49$.

Because of the uncertainties concerning the comparisons carried
out by Floyd and Bathe, it was decided to run the same example using
BEASY with $\nu = 0.4$ in order to compare results with ADINA. Results
were computed for two very simple meshes consisting of 20 and 31 quadratic

elements (giving 120 and 186 degrees of freedom respectively) as shown in figure 21a and 21b. Results were compared against the 69 elements (500 degrees of freedom) and 240 elements (1500 degrees of freedom) employed using ADINA. (The mesh with 491 elements not considered here was deemed to be badly conditioned and consequently giving no more accurate results than the 240 element mesh).

Figures 22a to 22c show the maximum principal stress, the minimum principal and the hoop stresses and compared results for the two finite element runs against the two boundary element runs.

It is interesting to note that the results obtained with BEASY exhibit a far better convergence for this problem than those obtained using the finite element program ADINA. The stresses along the line of interest are almost identical for the 20 and the 31 boundary element analysis. This shows that a relatively coarse mesh produces results of extraordinary accuracy using boundary elements and that the method does not experience any particular difficulty when studying reentry corners.

Can BEM Deal with Infinite Problems?

One of the most difficult problems to model using numerical techniques such as finite differences and finite elements is the case of problems whose domains extend to infinity. In BEM instead infinite or semi-infinite problems can be solved accurately as the influence solution is still valid to infinity. By contrast the 'domain' methods require truncation of the mesh at a certain distance and the application of some approximate boundary conditions there. Problems with infinite domain occur frequently in engineering in applications such as mining, soil mechanics, offshore and many others.

Figure 23 depicts a tension leg platform built for the North Sea. The problem under consideration is the study of the voltage and intensity around the platform due to the application of an impressed cathodic protection system. Figure 24 shows the boundary element mesh required for the analysis of the 'system'. The requirement was to solve Laplace's equation in the infinite domain of the seawater and the designer wished to know the current density on the cathode, which is the hull of the platform. Since the boundary of the seawater is the platform itself, the example was ideally suited to the boundary element method. There are

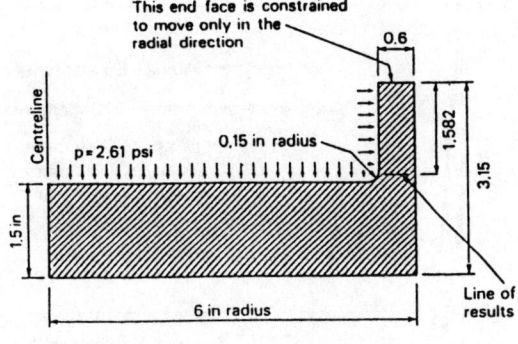

Fig. 20 Axi-symmetric pressure vessel section (Fig. 8 of [2])

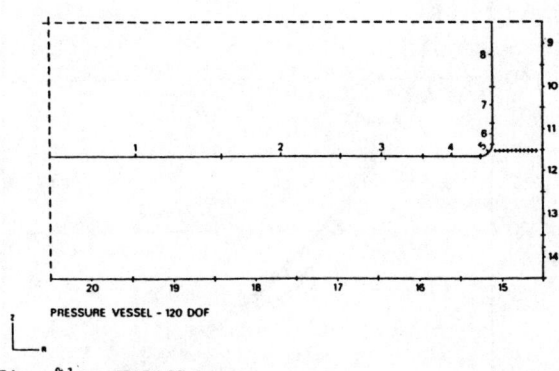

Fig. 21a BEASY 20 elements

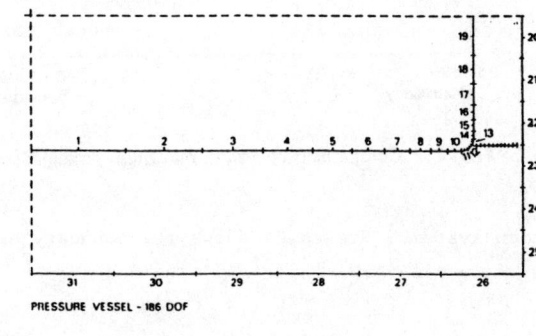

Fig. 21b BEASY 31 elements

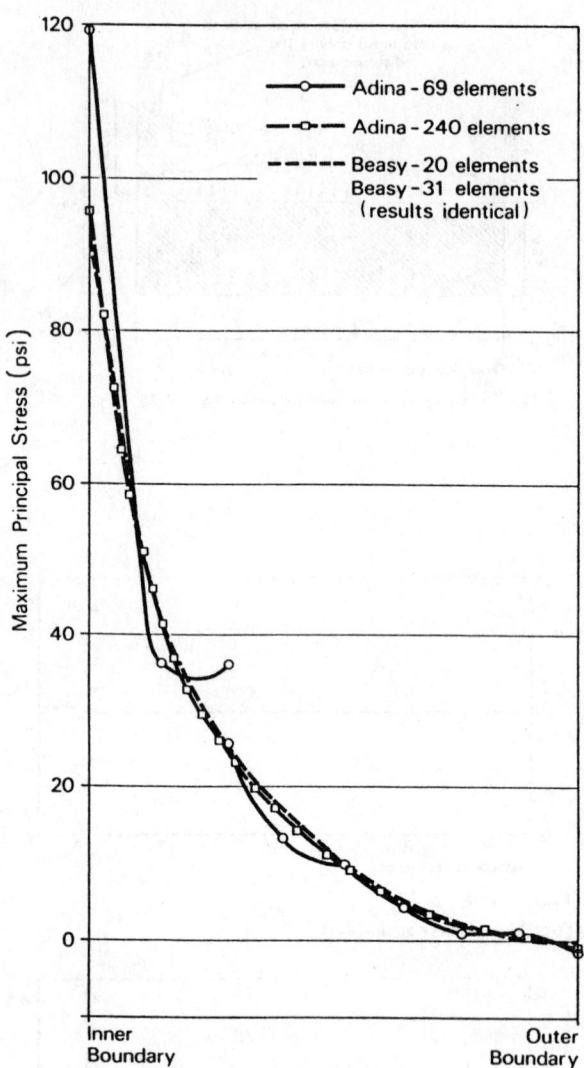

Fig. 22a Comparison of results-maximum principal stress

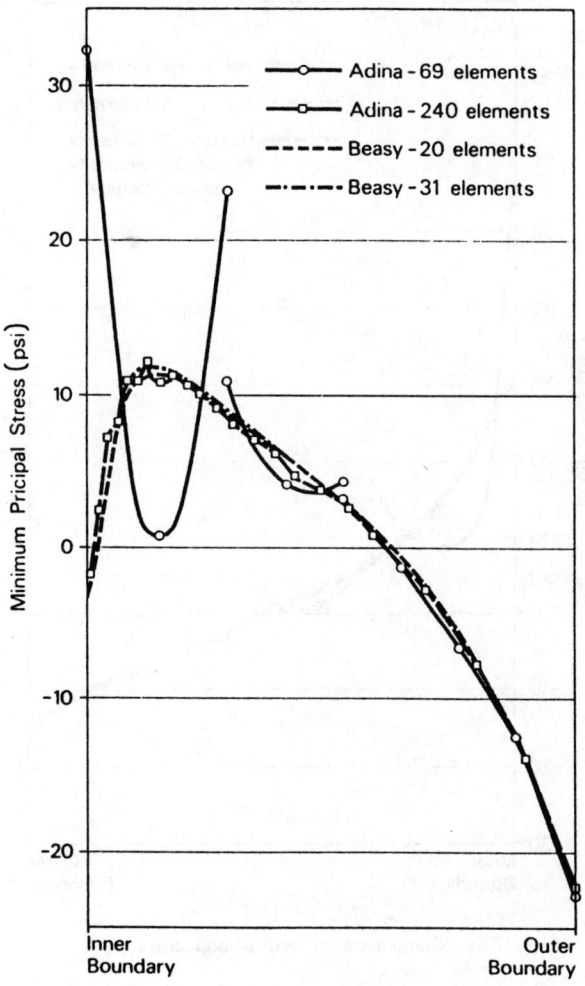

Fig. 22b Comparison of results-minimum principal stress

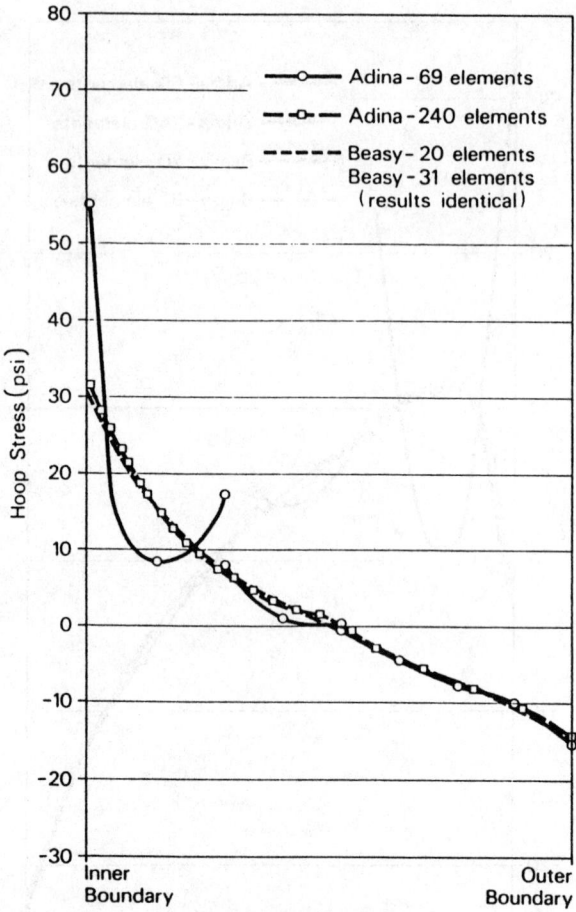

Fig. 22c Comparison of results-hoop stress

Fig 23 The Hutton Field Tension Leg Platform

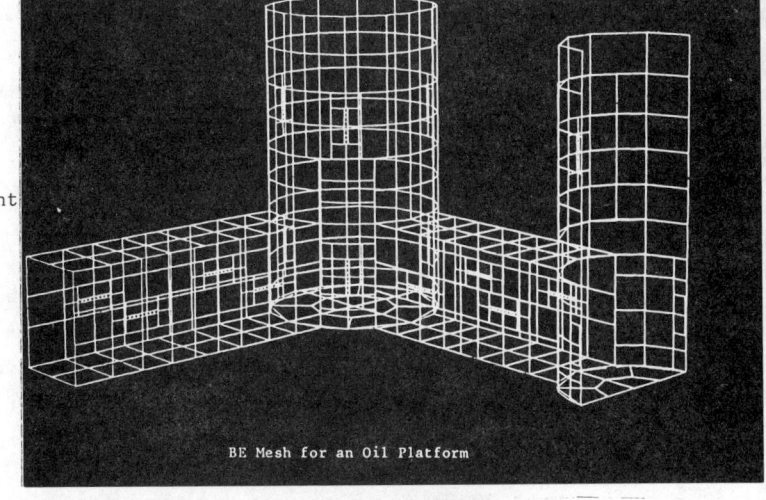

Fig 24 Boundary Element Mesh

Fig 25 Distribution of Currents

three planes of symmetry in this problem and the mesh shown in the
figure represents one quarter of the total problem. Two of these planes
are genuine planes of symmetry and the third plane is the actual sea
surface which enforces the boundary condition of zero flux and saves having
to put elements there. There were also some elements not shown in the
figure a large distance from the platform. These were put there to
enforce the boundary condition required by the designer of no flux at
a long distance from the platform and are not always needed. They could
be replaced by a subsidiary condition. Without these elements or special
subsidiary conditions, the BEM would automatically enforce a boundary
condition of potential equal to zero at infinity which is not what the
designer wanted in this case. It should also be noted that this auxiliary
mesh at a large distance from the structure is not connected to the mesh
on the platform itself. The smallest elements used in this problem had
dimensions 460 x 45 mm (on the anodes) and the largest has dimensions of
30 x 30 m (on the 'infinite' boundary). The problem was run using constant
elements in BEASY and was analysed using 653 elements. Several analyses were
performed as the designer not only wanted results for the complete system
but also required to know what would happen if various combinations of nodes
were switched off. The ability of the BEASY code to restart the analysis
at various intermediate points was very useful here.

Figure 25 shows a distribution of currents over the structure. Note the
position of the anodes. It is difficult to compare these results since
nothing similar has been produced using other numerical techniques.
The BEM is indeed the only feasible technique to analyse this type of
problem. The designers, however were satisfied with the solution and
have used them to design the cathodic protection system for the platform.

Although the previous example was ideal for being solved using
boundary elements, the method can also be applied to jacket type structures.
The analysis of an offshore structure like that shown in figure 26 where
the medium extends to infinity can be tackled in two stages. An overall
analysis of the structure can be performed using tube and anode elements
which would provide details of the current density and potential distributions along the members of the structures. This is an approximate

Fig 26 Jacket Type Structure Detail

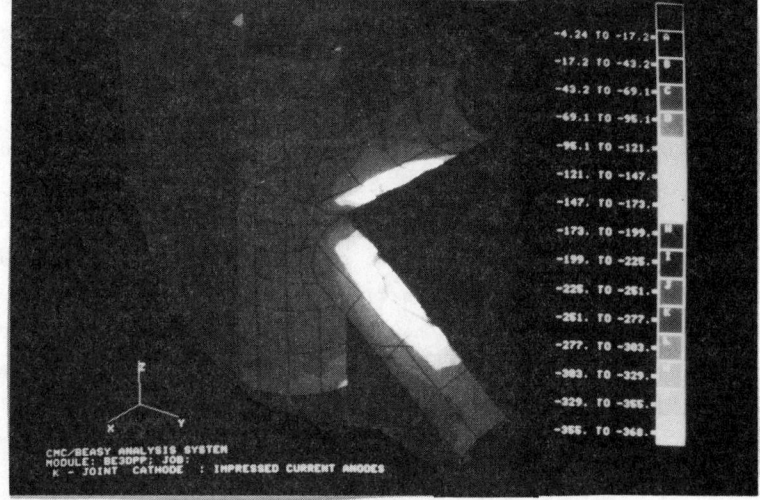

Fig 27 Discretization of a K Joint

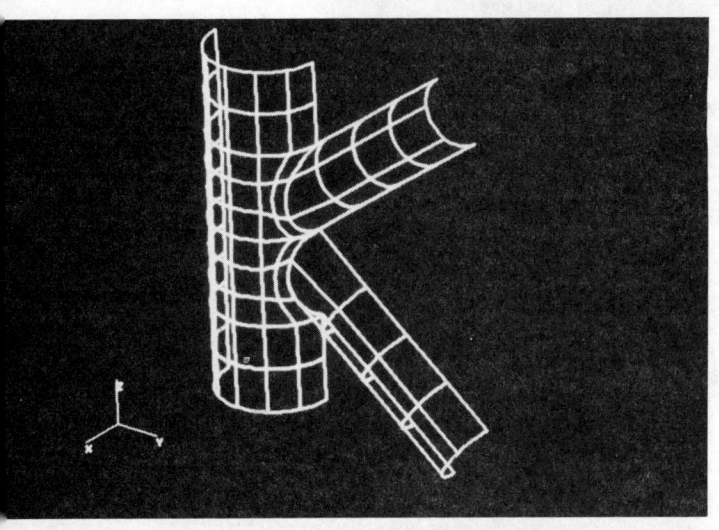

Fig 28 Intensity Contours on the K Joint

analysis as it reduces the members to one dimensional elements but it gives the designer details of the anode consumption rates and the degrees of protection provided, which can be used to assess the efficiency of the initial design. This 'global' analysis can be repeated for various configurations of anode systems to obtain the optimum design.

The overall or global analysis will not however produce full details of the shadow effects in individual joints, as the tube elements only provide the average potential and current densities at any point along the members (i.e. with no difference between front and back). Therefore special facilities have been provided in this system (BEASY - Cathodic) to zoom into individual joints using the values of the overall field computed in the 'global' model to provide the boundary conditions for the 'local' model. This local box model can now be used to compute a detailed analysis of the joint within the local box including shadow effects using generally curved surface elements.

Figure 27 shows the discretization of a K joint in which symmetry has been used and figure 28 some of current density contours obtained on the steel surface. The local box used is shown in more detail in figure 29 which illustrates more clearly the 'local box' used and the planes of symmetry which reduce the number of elements required. In this case the codes have been represented by line sources and a linear polarization curve has been applied on the steel surface of the joint. Figure 30 shows a contour plot of the predicted levels of potential on the surface. Non linear polarization curves can also easily be introduced as the non linearities will only affect the boundary conditions.

Has BEM Serious Advantages for the Designer?

As has already been mentioned design is by nature an interactive process, by which the engineer tries to find an optimum solution to the problems under a series of constraints. Although the process is highly dependent on the individual engineer (or the expert systems in the future?) the analytical codes at the core of the calculations are of primary importance.

BEM is an ideal technique for design purposes as it is available to the type of interactive process involved in CAD. Not only are meshes easily generated in BEM but what is perhaps more important, design changes can easily be incorporated as they usually do not require a complete remeshing.

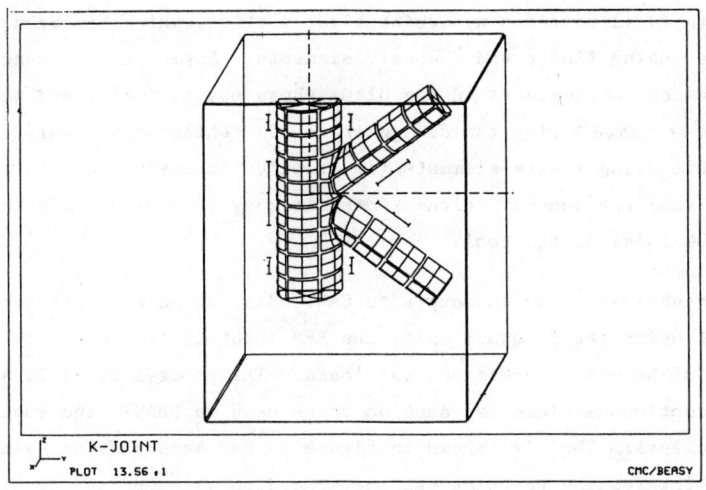

Fig. 29 Local Box showing discretization of a K joint

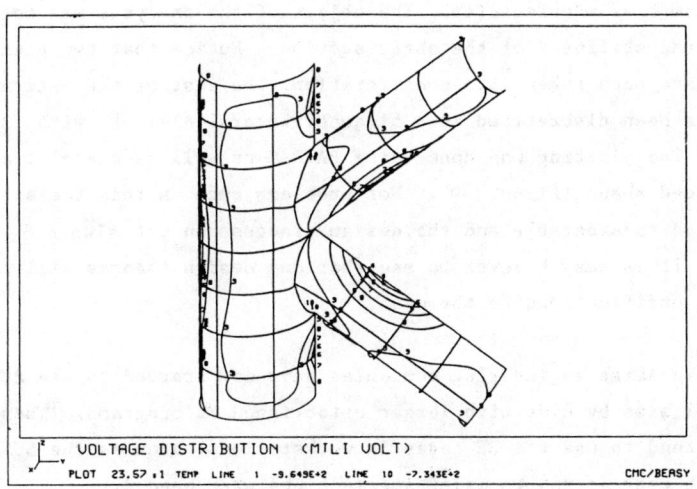

Fig. 30 Voltage Distribution Contour Plot on a K joint

Figure 31 illustrates a turbine blade section, which has been discretized using finite and boundary elements. Notice the presence of a series of coolingducts in the blade whose position,size and number need to be reviewed during the design process. While such a variation is difficult using finite elements (in practice it needs a new remeshing for every run) the boundary element mesh is easy to modify. BEM in this case is the ideal design tool.

BEM meshes are easy to couple to CAD codes, as once the structure is defined using the boundary only, the BEM input is finished with the exception of boundary conditions and loads. The process is still simpler when using discontinuous elements (such as those used in BEASY) and modern solid modellers. The mesh shown in figure 32 has been plotted using PATRAN and it is interesting to point out that the display shown in the figure (or its rotated views) can ensure the user that no surface element is missing. Notice how the use of discontinuous elements allows for a simpler meshing of the bearing cap surface. Figure 33 shows the displaced shape of the bearing cap and figure 34 the contours for displacements using PATRAN.

Figure 35 describes the mesh used to analyse a typical crankshaft for an automotive manufacturer. The object of the analysis was to calculate the stiffness of the shaft section. Notice that two planes of symmetry have been taken into consideration. The rest of the external surface has been discretized into 51 quadrilateral elements with 1377 unknowns. The plotting was done using PATRAN as well as the plot of the displaced shape (figure 36). For problems such as this the solution time may be considerable and the design process can not always be done interactively. It is easy however to see that any design changes will require only minor modifications to the mesh.

Certain large engineering companies have now started to use BEM as a design tool side by side with larger established BE programs. These companies tend to use the BE codes in workstations, such as the SUN or APOLLO with easy access to mainframe machines when needed.

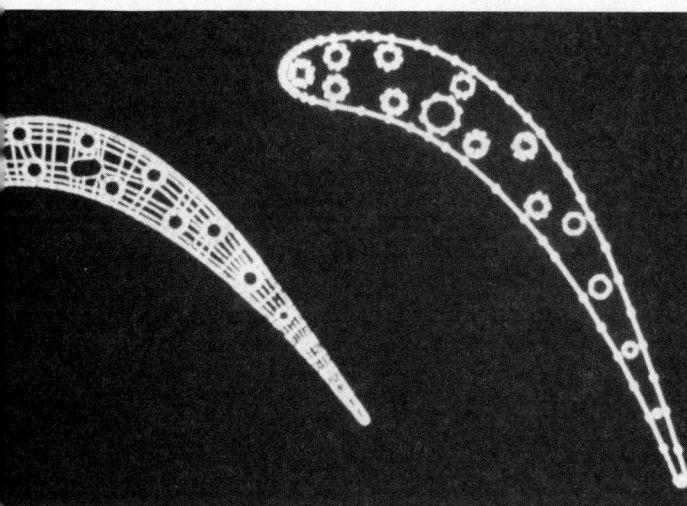

Fig 31 Turbine Blade Analysis using FEM and BEM

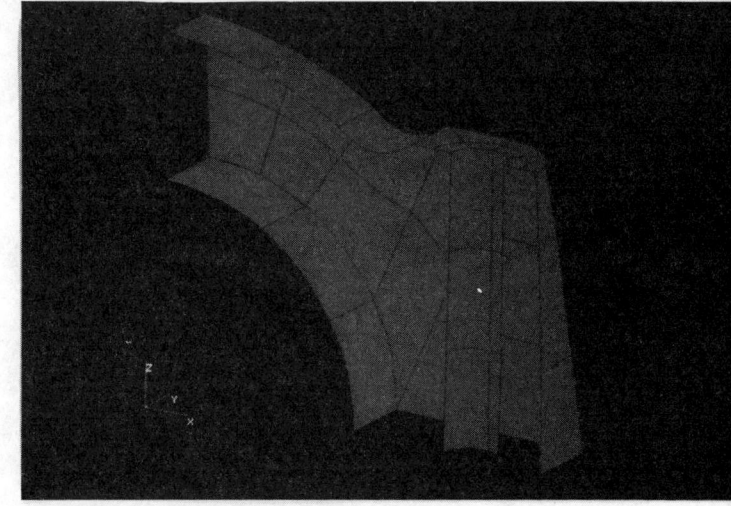

Fig 32 Bearing Cap Discretization

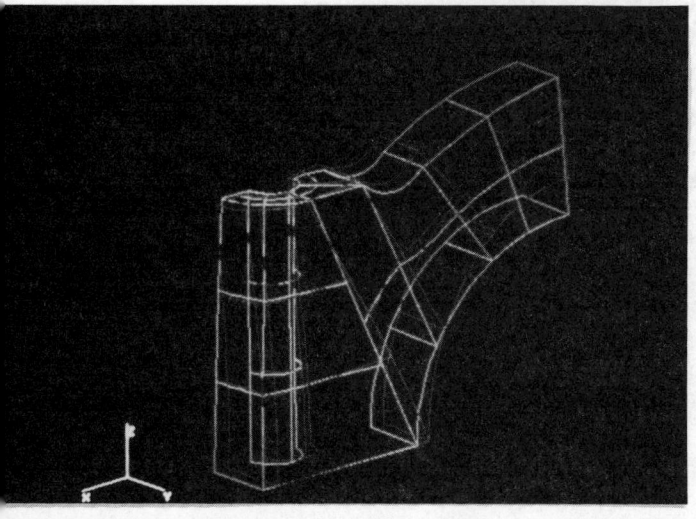

Fig 33 Displaced Shape for the Bearing Cap

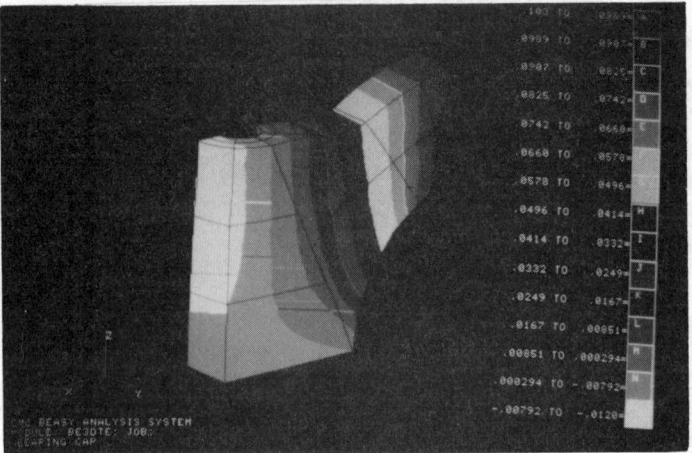

Fig 34
Displacement Contours for the Bearing Cap using PATRAN

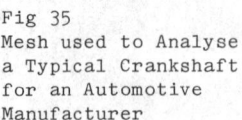

Fig 35
Mesh used to Analyse a Typical Crankshaft for an Automotive Manufacturer

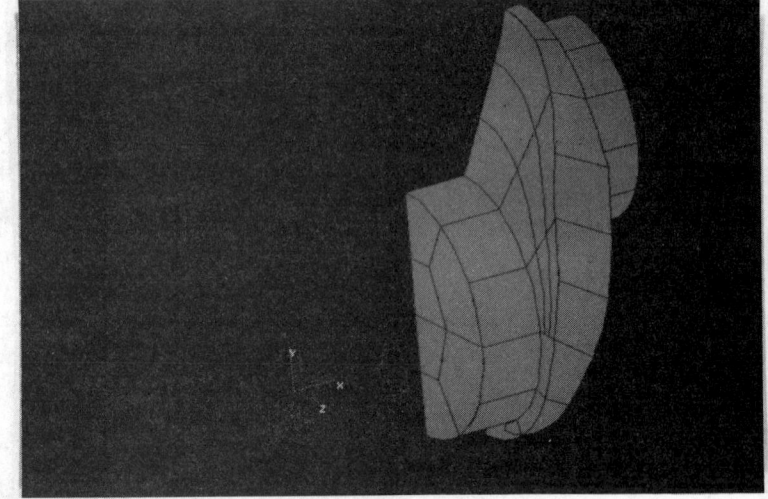

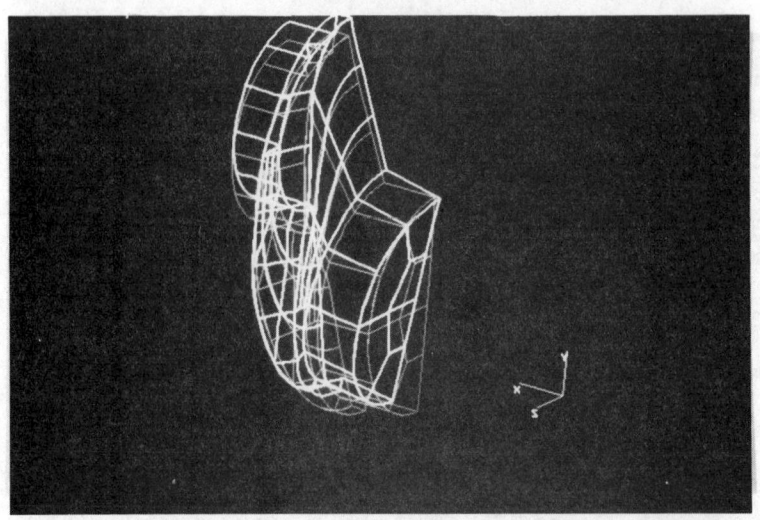

Fig 36
Displaced Plate Slot for the Crankshaft

What the BEASY Code Can Do

BEASY is at present the only fully developed boundary element system suitable for CAD work. The code has its own pre and postprocessing capabilities and can also be interfaced with packages such as PATRAN, SUPERTAB and FEMGEN/FEMVIEW. It runs in a wide variety of computers from CRAY to workstations such as the APOLLO and the SUN. The code has a library of two and three dimensional elements ranging from simple constant to quadratic curved elements. Although extra options in BEASY can solve a series of time dependent as well as non-linear problems, its most important characteristic in CAD is its ability to analyse thermal, electrostatics and general potential problems as well as elastostatics. In all cases elements are only required on the boundary and in the case of elastic problems body forces and thermal or centrifugal forces have also been taken to the boundary. The full range of linear options available in BEASY is shown in figure 37. Nonlinear options such as elastoplasticity, elastodynamics and time diffusion, although available in BEASY are not included in the figure as they are non-standard modules.

The great advantage of the BEASY code is that it simplifies the data required to run a problem, as only a simple - i.e. discontinuous - boundary mesh is required. This is the best feature of BEASY from the designers point of view. With computing costs still declining and engineer's time becoming more expensive the savings in engineer's time are far more significant than savings in machine time. Also, engineers welcome any advance that relieves them of the dreary task of data preparation and leaves them free to concentrate on more important tasks. Even more important is the fact that the analysis lies in the "critical" path in the design and production process and any tool which can shorten the "turnaround" time through the design office can bring forward the date of completion of the project. Although it is frequently stated that FEM generators can make that data as easy to prepare as BEM data, in the author's opinion, FEM mesh generation is still a problem. Industrial designers welcome the reduction in the order of mesh generation associated with BEM and appreciate the flexibility offered by the method, especially when using the type of discontinuous elements present in BEASY.

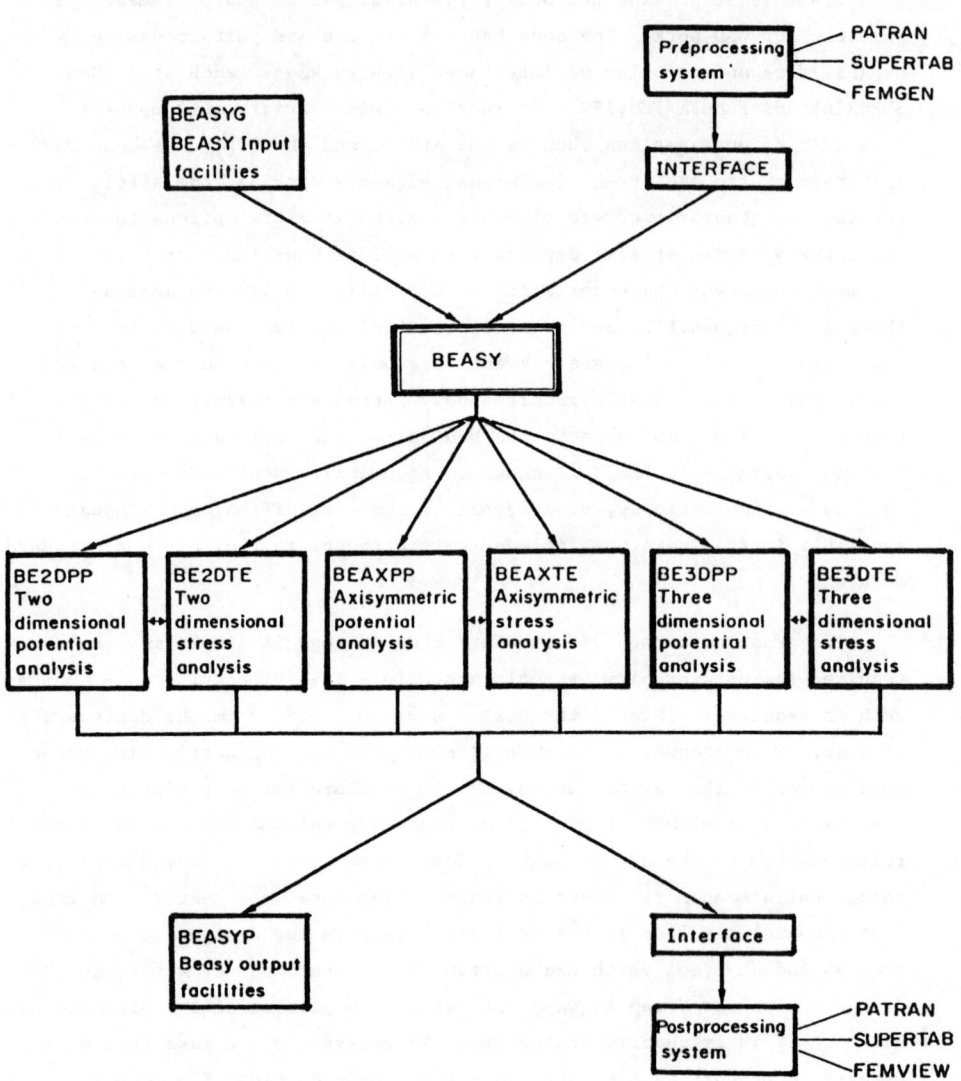

Fig. 37 BEASY System Modules

Our experience with practicing engineers indicates that designers are quick to appreciate the advantages of BEM and they can rapidly learn how to use the codes, without a deep understanding of the theory.

Where to go from here?

The future of BEM codes in engineering practice is promising and it will continue to be so provided that BEM practitioners do not alienate the engineering community by producing cumbersome or unreliable codes. Researchers expanding the technique should keep in mind the following points.

i) It is important that codes respond to the needs of the users, i.e. the practicing engineer and retain the simplicity of application, which is the main attraction of the BE codes. Sophisticated programs difficult to use will alienate the user who will then tend to focus on existing FE codes. The price of easy to use BE systems may be added mathematical or computational complexity for the developer. A good example of this is to try to produce nonlinear or time dependent codes which do not require internal cells. Further simplifications in the use of the codes may be obtained through the application of better programming techniques and the principles of Artificial Intelligence.

ii) Although better computational performance is needed by most 3D BEM codes, one should not sacrifice precision and accuracy of the solution in order to achieve computer time savings. Any attempt to reduce the quality of BE software will in the long run, damage future development. In recent years there has been a growing demand for reliable software with the result that many users are now becoming concerned with the unreliability of many FE codes, many of which are well accepted in engineering practice. Applying coarse numerical integration techniques for instance may result in large savings in computer time and give reasonable results in many cases. Unfortunately they will tend to produce nonconvergent results and solutions which can not fully represent conditions like rigid body movement.

iii) The development of more powerful hardware and in particular modern supercomputers favours the use of BE codes. These codes, because of the square character of their matrices or submatrices, are better conditioned for parallel or vector processing than FE codes which have banded matrices. Recent numerical tests carried out by the author on a CRAY machine suggests savings of 10 times or more in solution time when using vector processing. The widespread use of this type of machine will contribute to the propagation of BE systems.

iv) More research is needed into ways of solving different types of non-linear and time dependent problems using BEM. Although large strides have been made in the last few years (see for instance the Proceedings of the International Conferences on BEM 1 to 7), there is still a tendency to produce rather complex codes from the user's point of view, which include domain as well as boundary discretization. It is necessary to try to simplify the BEM formulation of these problems if at all possible.

v) A large programme of training is required to publicize the technique amongst practicing engineers. University courses for instance are still not geared to educate students in the fundamentals and uses of BEM. Short courses and special workshops could help to bring the method to the attention of those engineers who have already left University and were not exposed to the fundamentals of BEM. While several important books have recently appeared on the theoretical aspects of BEM, more applied publications are needed to reach the engineering community.

References and Bibliography

1. BREBBIA, C.A. "The Boundary Element Method for Engineers", CM Publications, Revised Printing, 1984.

2. BREBBIA, C.A., J. TELLES and L. WROBEL "Boundary Element Techniques - Theory and Applications in Engineering", Springer Verlag, Berlin and NY, 1984.

3. BREBBIA, C.A. (Ed.) "Progress in Boundary Element Methods, Vol.1" Pentech Press, London, 1981.

4. BREBBIA, C.A. (Ed.) "Progress in Boundary Element Methods, Vol.2" Pentech Press, London, 1982.

5. BREBBIA, C.A. (Ed.) "Topics in Boundary Element Research, Vol.1 Basic Principles and Applications" Springer Verlag, Berlin, NY, 1983.

6. BREBBIA, C.A. (Ed.) "Topics in Boundary Element Research, Vol.2 Time Dependent and Dynamics" Springer Verlag, Berlin, NY, 1985.

7. BREBBIA, C.A. (Ed.) "Recent Advances in Boundary Elements" Proceedings 1st International Conference on Boundary Elements, Southampton, Pentech Press, London, 1978.

8. BREBBIA, C.A. (Ed.) "New Developments in Boundary Element Methods" Proceedings of the 2nd Int. Conf. on Boundary Elements, Southampton, CML Publications 1980.

9. BREBBIA, C.A. (Ed.) "Boundary Element Methods" Proceedings of the 3rd International Conference on Boundary Element Methods, Springer Verlag, Berlin, NY, 1981.

10. BREBBIA, C.A. (Ed.) Boundary Elements in Engineering. Proceedings of the 4th Int. Conf. on Boundary Element Methods, Southampton Springer Verlag, Berlin, NY, 1982.

11. BREBBIA, C.A., T. FUTAGAMI & M. TANAKA (Eds.) Boundary Elements V Proceedings of the 5th Int. Conf. on Boundary Element Methods, Hiroshima, Japan, Springer Verlag, Berlin, NY, 1983.

12. BREBBIA, C.A. (Ed.) Boundary Elements VI. Proceedings of 6th Int. Conf. on Boundary Element Methods, held on board Queen Elizabeth 2, July 1984, Springer Verlag, Berlin, NY, 1984.

13. BREBBIA, C.A. & G. MAIER (Eds.) Boundary Elements VII. Proceedings of the 7th Int. Conf. on Boundary Element Methods. Villa Olmo, Italy, Springer Verlag, Berlin, NY, 1985.

14. FLOYD, C.G. "The Determination of Stresses using a Combined Theoretical and Experimental Analysis Approach", Proceedings of the 2nd Int. Conf. on Computational Methods and Experimental Measurements, held on board the Queen Elizabeth 2, June/July 1984, Springer Verlag, Berlin, NY, 1984.

15. SUSSMAN, T. and K.J. BATHE "Studies of Finite Element Procedure - as Mesh Selection" Computers and Structures, Vol.21, No.1/2, pp.257-264, 1985.

16. BREBBIA, C.A. and J. TREVELYAN "On the Accuracy and Convergence of B.E. Results for the Floyd Pressure Vessel Problem" Submitted for publication in Computers and Structures.

Chapter 3
SOME ENGINEERING SOFTWARE

BEASY: A BOUNDARY ELEMENT SYSTEM FOR STRUCTURAL ANALYSIS

C. A. Brebbia

Computational Mechanics Institute, Southampton, UK

ABSTRACT

This paper describes the structural analysis capabilities of the boundary element analysis system (BEASY) developed at the Computational Mechanics Centre, Southampton, England. The code is based on the boundary element technique and can be used to solve two-dimensional, axisymmetric and three-dimensional potential and elastostatics problems. Only the boundary needs to be discretized, thus reducing by one dimension the data required to run a problem. The code has excellent pre- and postprocessing facilities and can be interfaced to a range of well-known data generating and postprocessing systems.

The full text of this presentation was published in Niku-Lari A (ed.) *Structural Analysis Systems*, Vol. 1, pp. 49–60. Pergamon, 1986.

KYOKAI: A 2/D POTENTIAL AND ELASTICITY PROBLEM SOLVER ON SUPER-MINI

K. Onishi

*c/o Professor W. Wendland, Fachbereich Mathematik, Technische Hochshule Darmstadt,
Schloßgartenstr. 7, 6100 Darmstadt, FRG*

ABSTRACT

Recent development of the KYOKAI program is presented. The program is a small menu-driven command system for the solution of two-dimensional potential and thermoelasticity problems in zoned inhomogeneous materials. One of the FEM and BEM, or both, can be selected at the user's will. The system is the integration of pre-processing, analysis and post-processing. Small drafter and editor subsystems are incorporated in the pre-processor. The drafter is used for drawing the geometry of the domain in question and the editor is used for the modification of all the input data. The system can be quitted and then restarted at any stage during the execution in order to enable the stop-modify-go procedure. This is especially designed for the interactive use and for self-training of the computational methods on super-mini computers.

KEYWORDS

Finite element method; boundary element method; application program.

INTRODUCTION

This is the supplement to the paper by Kobayashi et al.(1985). The comprehensive description about the KYOKAI system was presented in Ohura et al. (1985). Here we shall focus on the interrelationship between subsystems and intermediate data files. Limitations on the size of the numerical model for corresponding boundary value problem will be described.

SUBSYSTEMS AND FILES

The interrelation between KYOKAI subsystems and data files is shown in Figure 1 in which subsystems are indicated by rectangles. On the extreme left, display and print devices are indicated by corresponding shapes.

BLOCK MESH subsystem defines the outline of the geometry by the combination of eight-noded quadrilaterals. Planar lines and quadratic curves are used. Boundary conditions are defined in this subsystem. Constant, linear and parabolic variations for the boundary values can be given. Each step in the model construction is monitored by a graphic display. Data for the geometry and boundary conditions are stored in the data file IG1. By the subsystem BCHN, the contents of IG1 can be changed and modified.

FINE MESH subsystem generates fine meshes of finite and boundary elements. Weighting factors for the relative spacing in the mesh subdivision must be specified only once by the user, and they are saved in IG2. The block mesh stored in IG1 is read and subdivided into a triangular mesh, and the mesh data are stored in IG3. Corresponding to the fine mesh, the boundary conditions defined already by the BLOCK MESH are transplanted on the boundary nodes. Some specific boundary conditions, which could not be introduced by the BLOCK MESH, can be given by this subsystem. Boundary data are stored in IG4. By the subsystem FCHN, the contents of IG2, IG3, IG4 can be changed and modified.

If some external forces are required for the problem in question, they are specified by the subsystem LOAD CONDITION. The conditions are defined to the fine mesh. The output data are stored in IG5, whose contents can be modified by FCHN.

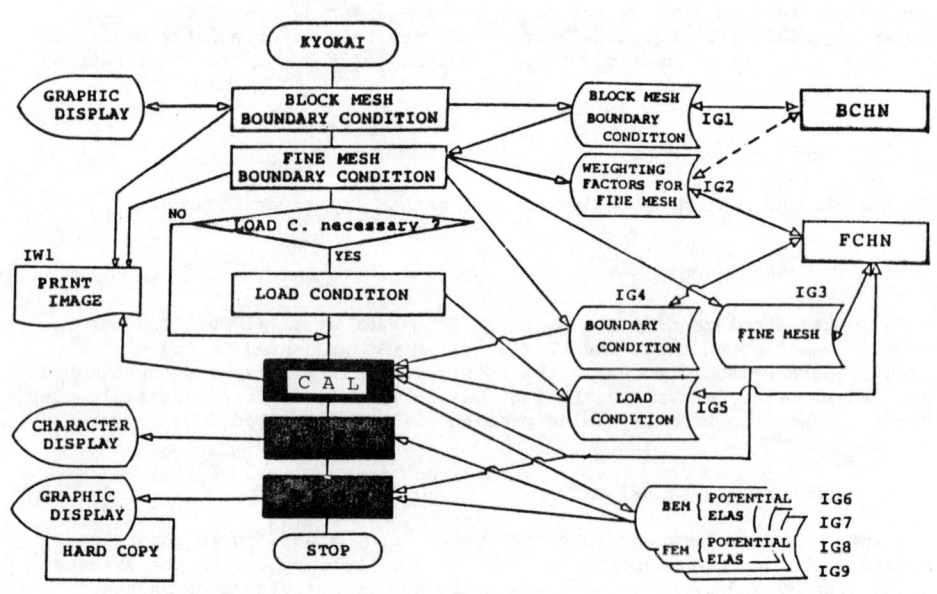

Fig. 1. KYOKAI subsystems and data files.

CAL subsystem is the equation solver. It reads IG3, IG4, IG5 and starts the
discretization of equations by either finite element method or boundary
element method. Numerical solutions are stored in one of the data files
IG6 - IG9 depending on the method of solution and on the type of the problem.
Suppose that a problem is solved first by the finite element method. Then
the command BEM resumes the solution of the same problem by the boundary
element method, and vice versa.

LIST subsystem displays calculated results in numbers. It reads one of the
output files IG6 - IG9 and shows only the desired part of the results on a
screen with appropriate headings and self-explanatory item names. PLOT subsystem draws calculated results in pictorial forms on a graphic display.
For the solution of potential problem, contours, curved surfaces and flux
vectors are plotted. For the solution of elasticity problem, displacement,
principal stresses, contours of the stress components and surface tractions
are plotted. The size and the position of the picture can be chosen arbitrary.
Figure 2 shows an output example of calculated temperature distribution in
a cylinder with cooling ducts.

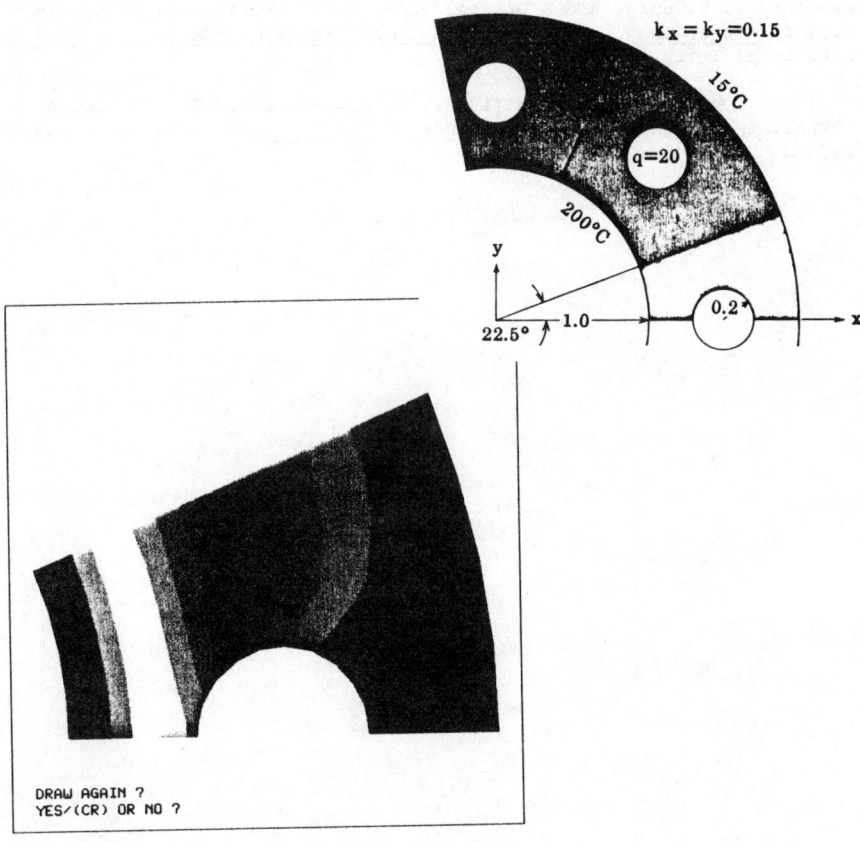

Fig. 2. Temperature distribution.

LIMITATIONS

Maximum numbers of finite elements and nodes are 75o and 5oo, respectively. Maximum numbers of boundary elements and nodes are 35o and 35o. The maximum number of non-zero elements in the coefficient matrix of constitutive linear equations is 5oooo in the course of Gauss elimination for the finite element method. These limitations can be relaxed by replacing corresponding array sizes in COMMON blocks in the main program and a DATA statement in the sourse program.

ACKNOWLEDGEMENT

This paper is written during the author's stay by Professor W. Wendland as a research fellow of Alexander von Humboldt Foundation.

REFERENCES

Kobayashi,K., Y.Ohura, and K.Onishi(1985). Small expert BEM.FEM system - KYOKAI. Proceedings 7th International Conference on BEM in Engineering, Lake Como, Italy, II, 11-3.

Ohura,Y., K.Obata, and K.Onishi(1985). An user-friendly BEM.FEM solver - KYOKAI. in A. Niku-Lari(Ed.), Structural Analysis. 1. Pergamon Press, Oxford.

MODULEF: A LIBRARY OF COMPUTER PROCEDURES FOR FINITE ELEMENT ANALYSIS

M. Bernadou, P. L. George, P. Laug and M. Vidrascu*

INRIA, Rocquencourt, B.P. 105, 78153 Le Chesnay, France

ABSTRACT

MODULEF is a general purpose finite element computer program developed by the MODULEF CLUB (see Appendix at the end of the paper).

Created by INRIA in 1974, this Club brings together French and foreign universities, and private or public industrial companies with the goal of designing and implementing a library of finite element modules.

Some existing module capabilities include solutions to:

- steady state or time-dependent, linear or nonlinear 2-D, 3-D and axis heat conduction problems

- static or dynamic linear or nonlinear 2-D, 3-D and axis elasticity problems

- elasticity problems for beams plates and shells

- fluid mechanics problems.

The various algorithms and software attributes such as modularity, portability and dynamic memory allocation make the MODULEF library a powerful tool for Research and Development.

Modular structures allow simple modifications and program additions.

Not only one, but several solutions to a problem can be readily implemented, and therefore the relative merits of various approaches can be easily assessed.

An interactive data generation system is available.

It is suitable for automatic mesh generation, generation of boundary conditions and generation of data needed for the main steps used in finite element computing programs.

---

*This is a collective work; it was carried out thanks to D. Begis, J. M. Boisserie, J. M. Crolet, A. Hassim, F. Hecht, P. Letallec, A. Perronnet, F. Pistre and D. Steer.

Furthermore, it has proved to be most useful for teaching purposes.

The full text of this article was published in:

Niku-Lari (ed.) Structural Analysis Systems, Vol.3, pp.153-73. Pergamon 1986.

MEF/MOSAIC: LINEAR, NONLINEAR FINITE ELEMENT CODE WITH INTERACTIVE GRAPHIC DISPLAY

J. F. Cochet

CSI/UTC, Centre de Recherches de Royallieu, B.P. 233, 60206 Compiègne, France

ABSTRACT

MEF is a finite element code to solve 2 or 3D problems of solid mechanics, fluid mechanics, heat transfer, electromagnetism, fluid structure coupling, piezoelectricity and others.

MEF was born in the nearly 80s. It is modular, easy to implement on microcomputers and includes interactive tools for graphic display (pre-post processor MOSAIC).

MEF is written in standard FORTRAN with precise programming norms and is easy to implement, modify, develop and to maintain. The program has been implemented on various 32-bit computers. MEF has been jointly developed by professors, engineers, graduate students from the University of Technology of Compiègne and from the University Laval of Quebec City.

MEF is industrialized and sold by the company C.S.I. (Compiègne Science Industrie).

Its purposes are to be used:

(a) as a very adaptative tool by companies which already own finite element codes.

(b) as teaching and research tool by institutes, universities and research centers.

(c) by companies which need specific developments in their fields.

The full text of this presentation was published in Niku-Lari (ed.) *Structural Analysis Systems*, Vol. 2, pp. 95-111.

NELIN3 – COMPUTER PROGRAM FOR NONLINEAR ANALYSIS OF SPACE FRAME AND CABLE STRUCTURE

J. Ožbolt

Faculty of Civil Engineering Zagreb, Yugoslavia
Gradjevinski Fakultet Zagreb, Kaciceva 26, 41000 Zagreb, Yugoslavia

ABSTRACT

FE computer program NELIN3 for materially and geometrically nonlinear static analysis is presented. The program is designed in such a way that it is possible to perform complete time history analysis of the structure. Material and geometric nonlinearity are taken into account, as well as creep, shrinkage and in time varying boundary conditions, so that construction sequence history can be simulated. The program has been originally developed for the analysis of reinforced concrete and steel structures on the Sirius--Victor microcomputer (512 kB, hard disk, 8087 coprocessor) and there is also a possibility for the graphic representation of some numerical results.

KEYWORDS

Finite element method; material nonlinearity; geometric nonlinearity; creep; shrinkage; construction history; microcomputer; restart.

INTRODUCTION

Quite often in the engineering practice a need for the nonlinear structural analysis arises. At present, there are several general purpose computer program in use throught the world which in many cases are not well suited for application in the solution of nonstandard problems. Also, these programs are most often installed on mainframe computers.
Here, a short description of the computer program NELIN3 for the nonlinear static analysis of space frame structures, consisting of combinations of frames and cables, will be given. The program has been written for the application on a personal microcomputer, and so far has been installed on the Sirius-Victor microcomputer.
In the analysis, material and geometric nonlinearity, system and load history in time (construction history or subsequent reconstruction or strengthening) and prestressing can be included. Elements may be of different age.

STRUCTURAL DISCRETIZATION AND MATERIAL MODEL

Spatial discretization of the structure is performed using the finite element method. As a finite element, the standard straight beam element is used. In the case of a cable element, only axial stiffness is considered and also compression is not allowed.
Material nonlinearity is taken into account by prescribing stress-strain diagrams for various materials (Ožbolt, 1981). At this stage of the program

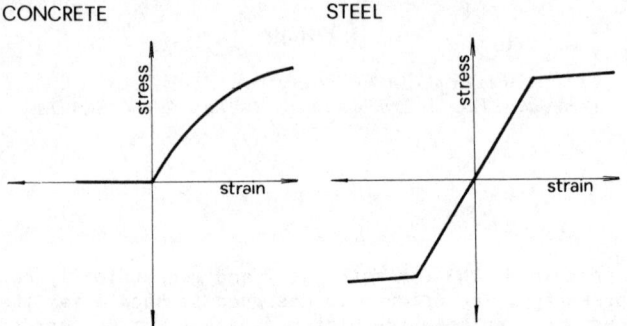

Fig. 1. Material laws for steel and concrete

development, hysteresis has not been accounted for. Since the interaction of shearing and normal stresses in the nonlinear domain is not known, these effects are superimposed as if being independent.
In a case of reinforced concrete elements, polygonal cross-section shape and the amount and the position of reinforcement need be input. From the known deformation, internal forces are determined using exact integration

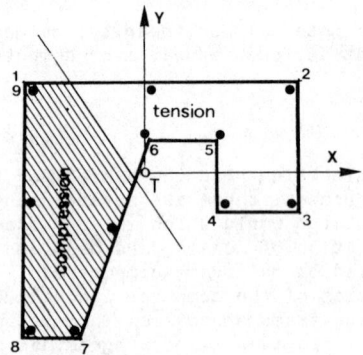

Fig. 2. Element cross-section

across the cross-section (Dvornik, 1974). In each element internal forces are determined at three cross-sections (at the element ends and in the middle) (Ožbolt, 1981). Equivalent nodal forces are determined using numerical integration along the element length and from the condition of equality of the work done by internal and external nodal forces.
Displacement field of the displacements perpendicular to the element axis is

assumed to be represented by a cubic parabola, and of the displacements along
the element axis by the quadratic parabola. Had the displacements along the
element axis been represented by a linear function, because of the material
nonlinearity it would be impossible to satisfy the equilibrium of internal
and external forces at the element level. Therefore it is necessary to add
another unknown to the unknown nodal displacements and that is the axial
displacement in the middle of the element. Prior to formation of the global
stiffness matrix this additional unknown is eliminated using static conden-
sation.

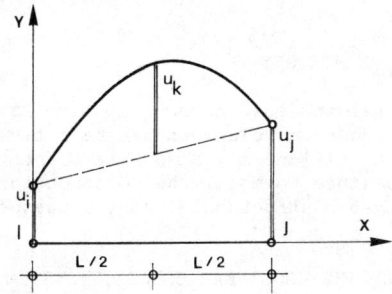

Fig. 3. Displacement field along the element axis

GEOMETRIC NONLINEARITY

Geometric nonlinearity is formulated using Update Lagrangian formulation. The
effect of large displacements and small strains is taken into account. During
the formation of the global stiffness matrix, geometric element stiffness is
considered. Equilibrium is satisfied on the deformed system. The global
instability is manifested as the stiffness matrix singularity. Local element
coordinate axes are defined by element end nodes and an extra orientation
node "K". As displacements are large, the local coordinate system is being
translated and rotated in space. The new orientation node position is
determined from the element end nodes displacements - translations and from
the average rotation about the element axis.

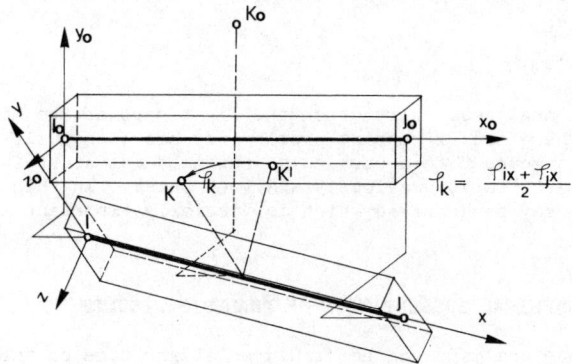

Fig. 4. Element orientation

CREEP AND SHRINKAGE OF CONCRETE

Creep and shrinkage for each element can be accounted for using and arbitrary creep and shrinkage law, depending upon the element age at the moment of assembling. Creep is determined according to the linear creep theory. Namely creep deformation is proportional to stress (Bažant, 1972). The creep law is defined by a series of partial creep factors for particular construction history phases and exploatation phases of the structure.
The shrinkage law is similarly defined (Bažant, 1972).

CONSTRUCTION HISTORY

Modelling of structural and load changes in time is provided for. Dependent of the construction sequence, elements can be activated and deactivated at any time. Boundary conditions may also be arbitrary varied in time.
In this way it is possible to model the construction sequence, assembling and disassembling of some structural parts, subsequent presstresing etc.

SOLUTION OF THE NONLINEAR EQUATION SYSTEM

The solution of the nonlinear equation system is usually the most time consuming part of the calculations. In program NELIN3 it is possible to use three methods:

- Constant stiffness method (Modified Newton-Raphson)
- Tangent stiffness method (Newton-Raphson)
- Conjugate gradient method.

Tangent stiffnes of each cross-section is determined by exact integration accross the cross-section and with regard of the present deformation. Element tangent stiffnes matrix is formed in each iteration step using numerical integration of cross-section stiffnesses along the element length.
The solution of the equation system using the conjugate gradient method does not require the explicit formation of the global matrix with. The multiplication of the stiffness matrix with the residual vector is realized by adding contributions of each element. In the nonlinear case, the method compares favourably to standard method only when the number of equations is relatively small. The efficiency could be significantly improved by the use of the generalised conjugate gradient method (Dvornik, 1978).

RESTART

In practical analyses, very often the first part of the history has to be repeated. Therefore, a restart possibility has been built into the program NELIN3. This means that a particular run need not start from the beginning but from an arbitrary, previously analysed phase. In this way large computer time savings may be achieved which is specially important in microcomputer applications.

GRAPHICAL PRESENTATION OF ANALYSIS RESULTS

Results of the analysis can be displayed at any time on the computer display or be printed on the matrix printer. A perspective view of the structural model can be obtained as well as of the displacements in the blown-up scale.

APPLICATION OF THE PROGRAM

The program has been tested on several examples. Here, the results of two examples are presented.
In the first, a curved cantilever beam is loaded with the vertical force P. Material behaviour is taken as linear elastic, but displacements due to the vertical force are large and geometric nonlinearity had to be taken into account.

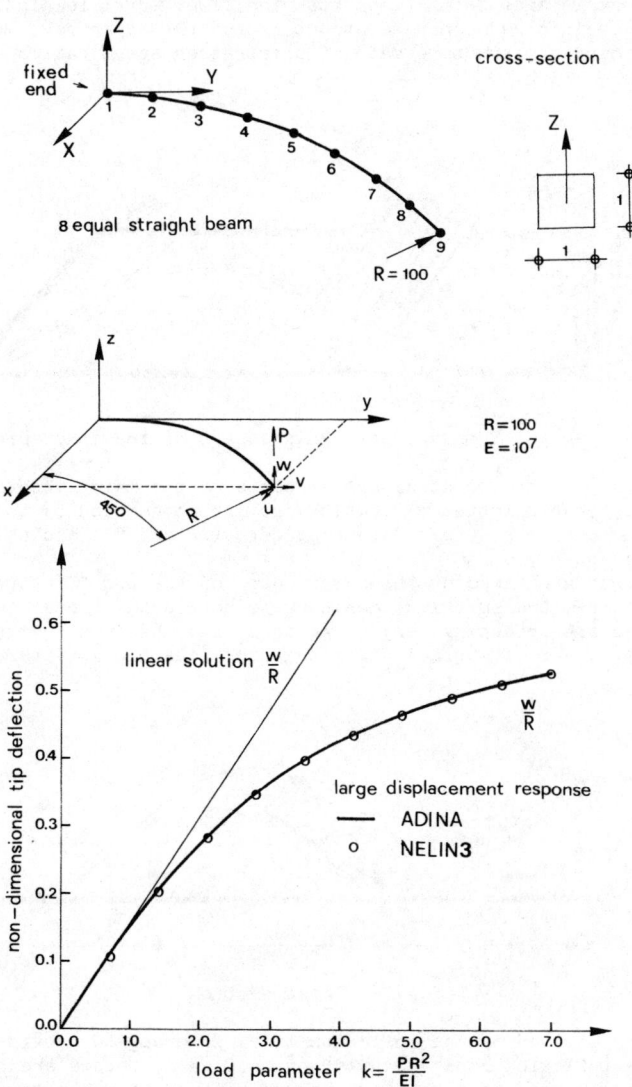

Fig. 5. Three-dimensional large deflection analysis of a 45-degree circular bend

A curved beam has been modeled using the beam element idealization. The results of the analysis obtained by the use of program NELIN3 are compared with the results obtained using the computer program ADINA (Bathe, Bolourchi, 1979). As it is shown, results of both analyses are in good agreement.
In the second example the results of the numerical analysis using program NELIN3 are compared with the measurements carried out on a swimming-pool structure hall in Split (Yugoslavia) which has been made on occasion of the Mediteranian Games in 1979. (Dvornik, Ožbolt, and Fejzo, 1986).
The structure is made up of two parts. The lower part is a partially prestressed reinforced concrete structure and the upper part is a roof constructed as a hanging structure made of prestressed steel cables.

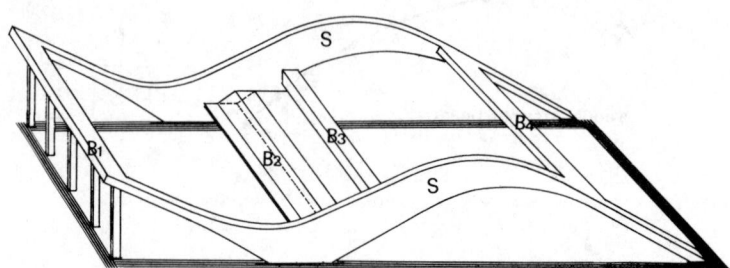

Fig. 6. The concrete (lower) part of the structure

The concrete part of the structure consists of two main girders shaped as a horizontally placed letter "S", which are interconnected by two transversal very slender prestressed grandstand girders (B2 and B3) and by two transversal girders (B1 and B4).
The roof structure hangs on the upper edges of the two "S" girders and of the beams B1 and B4, transmitting considerable horizontal forces to its supports. These forces are producing very large torsional and bending moments in the "S" girders. The roof surface has a negative Gaussian curvature and consists

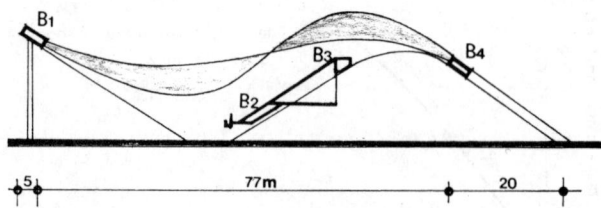

Fig. 7. Cross section

of two parts divided by the inflexion line. In order to provide enough stiffness in the inflexion zone, much larger roof cable forces are needed then in the rest of the roof, and this has rather unfavourable consequences for the concrete part of the structure. Namely, behaviour of the structure has been monitored and transversal girders with the span of 64 meters deflected by as much as 0,35-0,40 meters and numerous cracks opened up to 0,40 mm wide. A system of torsion induced cracks appeared on the "S" girders and the distance

between the girders decreased by cca. 0,40 meters. Due to this, the roof
structure lost the larger part of its prestressing forces and very large
deflections occured.
The global behaviour of the structure has been numerically interpreted using
the program NELIN3.
In the analysis it had to be taken into account:

- Material nonlinearity (for the reinforced part of the structure)
- Geometric nonlinearity (for the roof structure)
- Interaction of the reinforced concrete part and the roof part
  of the structure
  Here, the pronounced ill-conditioning of the problem had to be
  dealth with (the same model consists of very stiff "S" -girder
  elements as well as very flexible steel cable elements)
- Creep and shrinkage of concrete.

There exists a plane of structural and load symmetry in the structure and
therefore a model of only a half of the structure has been prepared.
The model consistes of 81 beam elements (concrete part of the structure) and
102 cable type elements (roof structure). As no adequate model for the

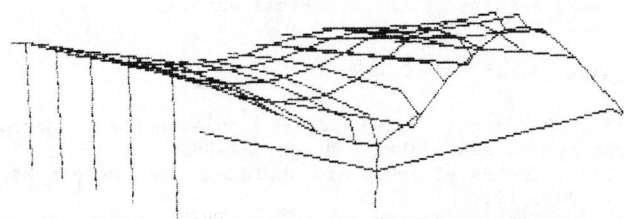

Fig. 8. Model of the structure

force acting on a box cross-section could be found in the literature, torsional
stiffness has been treated according to the guidelines of the Euroepan
Committee for Concrete. Bending stiffness has been determined using the
integral moment-curvature diagram (Monnier, 1970), and in the later stage of
the program development, the bending stiffness has been determined automatically by inputting the stress-strain diagrams for concrete and steel.
Numerical results obtained using program NELIN3 are in very good agreement
with data obtained during the monitoring of the structural behaviour.

CONCLUSION

A computer program NELIN3 for the nonlinear analysis of space frame structures
has been developed and installed on a microcomputer. The program is designed
in a way that it is possible to perform a complete history analysis of the
structure. Material and geometric nonlinearity as well as creep and shrinkage
of concrete and changes of the structural system in time can be accounted
for. Interruption and restart of the program at an arbitrary moment of the
analysis are possible.

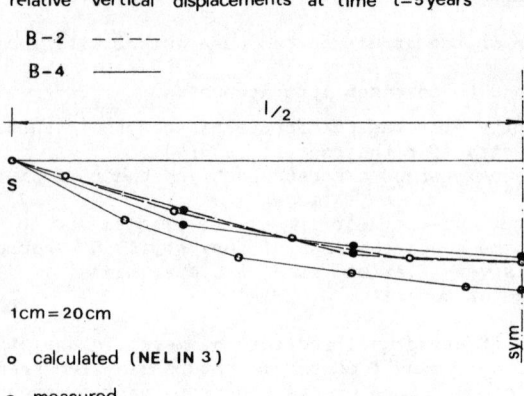

Fig. 9. Data obtained during the monitoring and
results of the numerical analysis

REFERENCES

Bathe, K. J., and S. Bolourchi. (1979). Large displacements of three-
dimensional beam structures. IJNME, 14, pp.961-986.
Bažant, Z. P. (1972). Theory of creep and shrinkage in concrete structures.
Mechanics Today, Vol. II.
Dvornik, J. (1974). Optimalno dimenzioniranje armirano-betonskih presjeka.
Disertacija. Zagreb. Yugoslavia.
Dvornik, J. (1978). Generalization of the CG method applied to linear and
nonlinear problems. Trends in computerized structural analysis and
syntesis. NASA Langely Research Center, USA. Chap. 5, pp. 217-225.
Dvornik, J., Ožbolt, J., and R. Fejzo, (1986).
Strengthening of the swimming-pool hall structure.
International Conference on Computer application in concrete. Singapore.
Monnier, Th. (1970). The moment-curvature relation of reinforced concrete.
Heron, No. 2. Nederland.
Ožbolt, J. (1981). Plasticity creep and shrinkage of concrete in one
dimensional numeric stress-strain analysis. TNO Rapport, BI-81-19/62.1.1110.
Rijswijk,Nederland.

NISA: THE PERFORMANCE OF NISA, A GENERAL PURPOSE FINITE ELEMENT PROGRAM FOR STATIC, DYNAMIC AND HEAT TRANSFER (LINEAR & NONLINEAR) ANALYSIS

Kant S. Kothawala

Engineering Mechanics Research Corporation, 1707 W. Big Beaver, Troy, MI 48084, USA

ABSTRACT

NISA is a large scale, general purpose, finite element program for analyzing actual problems encountered in civil, mechanical and offshore engineering environments as well as in the automotive, aerospace, nuclear, and many other industries.

KEYWORDS

Linear and nonlinear, static, dynamic, heat transfer, fatigue, structural optimization, shape optimization, fluid flow, electromagnetic, finite elements, structural analysis, CAD/CAM and CAE.

INTRODUCTION

A world leader in finite element technology, EMRC has improved the performance of the NISA finite element program significantly to satisfy the requirements of engineers worldwide. The performance is measured in terms of speed (400% increase), capabilities (50% increase in program size/capabilities offered), and user-friendliness (complete on-line documentation, automatic generation features, etc).

The NISA family of programs includes NISA-Static, NISA-Dynamic, NISA-Heat Transfer, NISA-Composite, ENDURE-a fatigue analysis program, NISOPT-a shape and structural optimization program, DISPLAY-an interactive graphics pre-processing program and DISPLAY II-a color and interactive graphics post-processing program.

NISA has evolved over the last decade into one of the most powerful, accurate and comprehensive large scale finite element programs available in the 80's. The NISA program today is used by hundreds of organizations spanning the full range of industry, research, and educational institutions.

The NISA family of programs encompass an assortment of general and special purpose programs for the design and analysis of structural and non-structural problems. These programs reflect the latest state-of-the-art techniques. NISA has been recognized as one of the most powerful and complete general purpose finite element programs.

The large and diverse finite element library coupled with the pre- and post-processors, interactive and color graphics capabilities, extensive analysis features, give the user the ability to take the product from concept to complete design.

NISA is a general purpose, finite element program for analyzing actual problems encountered in civil, mechanical and offshore engineering environments, and in the automotive, aerospace and other industries. Analysis capabilities include Linear and Nonlinear Static, Dynamic, Steady State and Transient Heat Transfer, 3D Fluid, and Electromagnetic. Element types available include isoparametric linear, parabolic, cubic, linear parabolic, linear parabolic cubic, etc. for plane stress, plane strain, axisymmetric (with symmetric or unsymmetric loading), general shells, solids, beams, spars, springs, mass elements and rigid elements. The extensive library of elements in NISA along with structural optimization (NISAOPT) allows one to design and analyze any size or shape of the structure.

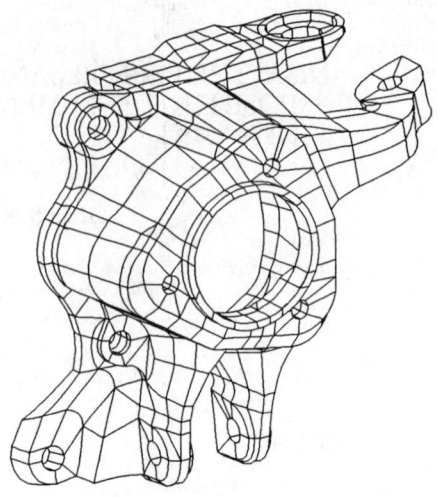

NISA MODEL OF AN AUTOMOTIVE KNUCKLE
575 20 NODE ISOPARAMETRIC SOLID ELEMENTS

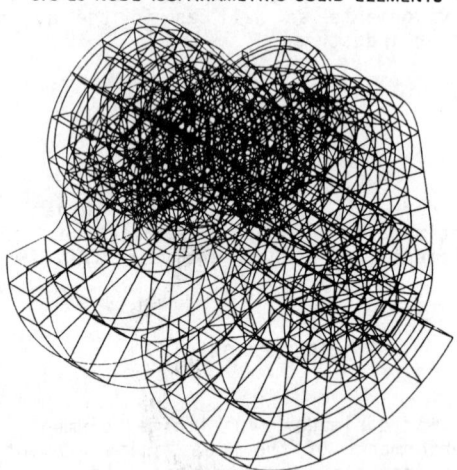

NISA Finite Element model of one throw of a four cylinder crank shaft using 20 node isoparametric solids

NISA offers the advantages of simplified data preparation in free format, comprehensive graphics capabilities for creation, automatic generation and verification of finite element models and numerous output options. NISA has been designed to be machine independent and can be implemented on most computers with minimum effort.

NISA and DISPLAY, an interactive color graphics pre- and post-processing program, have already been successfully implemented on the various mainframe computer systems of AMDAHL, BURROUGHS, CDC, CRAY, HITACHI, HONEYWELL, IBM, UNIVAC, and the superminis, minis, micros, and personal computers of ALPHA-MICRO, APOLLO, CIMLINC, CONCURRENT, DATA-GENERAL, ELXSI, ENCORE, EDGE, GOULD SEL, HARRIS, HP1000, HP A.600, HP A.700, HP A.900, HP9000, IBM PC XT/AT (AT & T, COMPAQ, ETC.) PRIME, SUN MICROSYSTEMS, VAX, and others. EMRC is in the process of implementing NISA/DISPLAY on ALLIANT, CONVEX, ETA and other computer systems.

NISA/DISPLAY is interfaced with the pre- and post-processors of the following CAD/CAM systems: APPLICON, AUTO-TROL, ANVIL 4000, CADAM, CAEDS, CALMA, DISPLAY II, GERBER, GRAFTEK, INTERGRAPH, MOVIE.BYU, PATRAN-G, SUPERTAB, etc.

The VECTOR PROCESSING capabilities of the CRAY and HP systems have been used to greatly speed up equation solving and eigenvalue extraction. NISA has PARALLEL PROCESSING capabilities on computers such as ELXSI, ENCORE, etc. with multiple CPUs and can be implemented easily on any other parallel processing computer with multiple CPUs.

NISA ANALYSES CAPABILITIES

As with any other large scale general purpose finite element program, NISA has automatic reduction and renumbering of wavefront transparent to the user. NISA has the most extensive data checking and pre- and post-processing capabilities.

The DISPLAY II program of EMRC can do all the data preparation from scratch. The pre- and post-processing capabilities available in DISPLAY II give tremendous flexibility in generating the entire model and studying the results through post-processing including color light shading and animation.

NISA gives extensive facilities to compute average nodal point stresses, node point stresses at element level, gauss point stresses and all failure criteria such as Von Mises, Tresca, octrahedra and Tsai-Wu (for composite). The program also gives a summary of results for the fifty largest stresses. The user may choose to suppress the printing of the summary of results with selected output.

DYNAMIC ANALYSIS

NISA has outstanding capabilities to solve a wide variety of problems encountered in dynamic analysis. It uses the wavefront technique with the obvious advantage that it can handle large problems efficiently and accurately. Model solution methods as well as direct integration can be used for any dynamic analysis.

Eigenvalue and Eigenvector Extraction Techniques:

NISA provides several different solution techniques to find the natural frequencies and corresponding mode shapes of a given structure. All three of these techniques are well proven as very efficient and accurate to solve for a large number of frequencies and mode shapes.

Subspace Iteration:

This method is an out-of-core solution technique. It is used to extract the natural frequencies which lie between zero and the upper frequency limit as specified by the user. This technique is very efficient to extract the first few natural frequencies.

Inverse Iteration with Sturm Sequence:

This method can be used to extract eigenvalues within an arbitrary range bounded by the cut off frequencies supplied by the user. A user specified number of frequencies at the low end of this range is calculated. The method first isolates each eigenvalue by means of Sturm sequence based bisection process and then applies inverse iteration technique.

QR - Householder

This method is an in-core solution technique which is very fast for small size problems. It gives all eigenvalues and eigenvectors of a given structure. For large size problems, this method can be used in conjunction with Guyan reduction.

Shock Spectrum Generation:

NISA Shock Spectrum Generation program may be used for the computation of shock spectra from raw time history records (earthquake records) for different damping values. This may be used as an aid to NISA Shock Spectrum Analysis. Plots of shock spectra thus generated can also be obtained.

Frequency Response Analysis:

NISA Frequency Response Analysis can compute the steady state response of a structure to harmonic loads defined by amplitude and phase spectra. Amplitude and phase spectra of displacements, velocities and accelerations may be obtained in the form of tables and plots. Displacements and stresses can also be printed at user specified frequencies and phase lags.

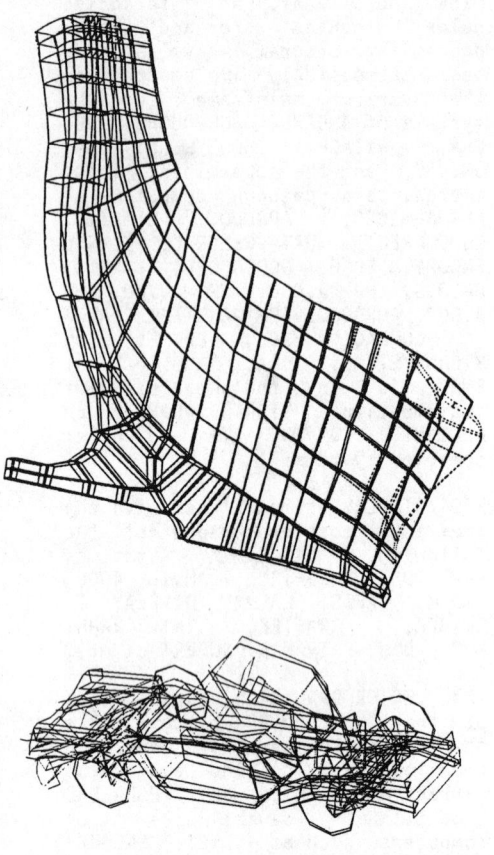

Random Vibration Analysis:

NISA Random Vibration Analysis can be used to obtain the statistical parameters of response quantities such as power spectral densities (PSD) and mean square values of structures subjected to a stationary random loading or ground excitation (such as earthquake, etc.). Plots of PSD's of displacement, velocities and accelerations can be obtained. Computation of cross spectral densities of displacements, velocities, accelerations and stresses and the mean square values of stress velocities is currently under development.

DYNAMICS - OUTPUT PRINTED AND GRAPHICS

Same as in STATIC plus the following additional features

* Eigenvalue (natural frequencies) and Eigenvectors (mode shapes)

* Node point displacement, velocity and acceleration vs. time

* Mode shape plotting

* Stresses and displacements printed at specified time intervals

* Displacement and stress contour plotting at specified time intervals

* Deformed structure plotting at specified time intervals

* Stress survey plots for composite shells at specified time intervals

Random Vibration

RMS displacements, velocities, accelerations and stresses

Printout and/or plots of PSD (power spectral density) of displacement, velocity and acceleration vs. frequency of selected degrees of freedom

Frequency Response:

Steady state displacements, velocities, accelerations and phase angles at specified frequencies of excitation

Plots of actual displacements, velocities, accelerations and stresses specified phase angles and frequencies

Plots of frequency vs. phase angle, frequency vs. steady state displacements, velocities and accelerations

Shock Spectrum:

Generalized displacements, velocities and accelerations

Option to print RMS (Square root of the sum of the squares of the modal responses), PRMS (Maximum absolute modal response added to RMS of remaining modal responses) and PEAK (sum of absolute values of modal responses) responses for displacements, velocities and acceleration

Option to print modal components of RMS, PRMS and PEAK stresses of the elements of the finite element model for the participating modes.

SHOCK SPECTRUM ANALYSIS OF A THREE-STORY FRAME

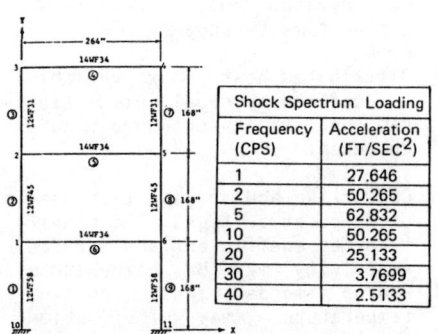

Shock Spectrum Loading	
Frequency (CPS)	Acceleration (FT/SEC^2)
1	27.646
2	50.265
5	62.832
10	50.265
20	25.133
30	3.7699
40	2.5133

Elastic Modulus = 30×10^6 psi
Poisson's Ratio = 0.3
Mass Density = .000735 lb-sec^2/in^4

	RMS DISPLACEMENTS (GLOBAL COMPONENTS)					
NODE	UX		UY		ROTZ	
	NISA	STRUDL	NISA	STRUDL	NISA	STRUDL
1 and 6	.267390	.267404	.002359	.002383	.001877	.001870
2 and 5	.680360	.680384	.004031	.004083	.001632	.001626
3 and 4	.959571	.959598	.004743	.004837	.000702	.000739

Option to print RMS, PRMS and PEAK stresses of the elements of the finite element.

Option to print modal components of RMS, PRMS and PEAK stresses of the elements of the finite element model for the participating modes.

NISA/HEAT TRANSFER

NISA can be used for both the steady state and the transient heat transfer analysis. Steady state heat transfer problems are analyzed with NISA in 1D, 2D and 3D configurations using the wavefront method of solution. Transient heat transfer problems are solved by using one parameter family of time iteration techniques. Both material properties and boundary conditions can be temperature and/or time dependent. Output temperatures can be stored for subsequent thermal stress analysis. The heat transfer capabilities of NISA can also be used for other field problems governed by the diffusive types of governing equation.

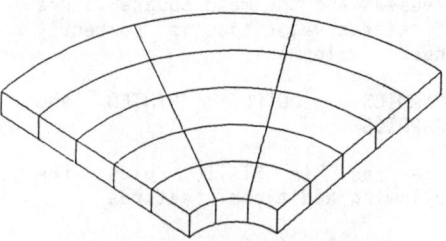

ONE QUARTER OF THE CIRCULAR DISC MODELLED USING 15 TWENTY NODE PARABOLIC SOLID ELEMENTS

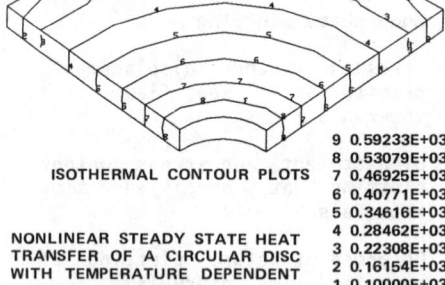

ISOTHERMAL CONTOUR PLOTS

NONLINEAR STEADY STATE HEAT TRANSFER OF A CIRCULAR DISC WITH TEMPERATURE DEPENDENT THERMAL CONDUCTIVITY.

```
9  0.59233E+03
8  0.53079E+03
7  0.46925E+03
6  0.40771E+03
5  0.34616E+03
4  0.28462E+03
3  0.22308E+03
2  0.16154E+03
1  0.10000E+03
```

STEADY STATE OR TRANSIENT THERMAL ANALYSIS

* Concentrated nodal heat flows can be time dependent

* Distributed heat flux variable as defined by nodal intensities can be time and/or temperature dependent

* Convective and/or radiative heat exchange at surfaces. Film heat transfer coefficient and surface emissivity may be temperature and/or time dependent. Ambient temperature may be time dependent

* Material properties (Thermal Conductivity Specific Heat, Density) may be temperature dependent. Temperature dependent properties can be a polynomial function of temperature or a piecewise linear interpolation

* Thermal conductivities can be orthotropic. Directions of orthotropy may coincide with global coordinates or be specified independently for each node or element.

* A general curved shell element is available. Heat flux and/or convection coefficients can be specified independently on the top and bottom surface of the shell, as well as on any of the inplane edges. The shell theory allows a linear temperature gradient through the thickness, as well as isoparametric variations of temperature in the inplane directions.

* Internal heat generation (average or distributed) can vary with time and/or temperature.

* Specified nodal temperatures may be time dependent

* In transient analysis, user may use variable time step size or define his own time step sizes.

* Nodal temperature-time history plots and temperature contour plots at specified time steps can be created.

OUTPUT FEATURES

* Nodal temperatures - average value at the midsurface of the shell elements

* Linear temperature gradients at the nodes of shell elements

* Heat values at the element nodes

* Heat values at the nodes with specified nodal temperatures

* Convective heat calculations for the element faces. Positive values indicate heat into the structure

* Isothermal contour plots

* Nodal temperatures at selected time intervals

* Isothermal contour plots at selected time intervals

* Temperature vs. Time plots for selected nodes

NISA - COMPOSITES

NISA - Composites, which is a part of NISA, is tailored specifically for accurate and efficient analysis of a wide range of composite structures. The following list includes a few of the many composite analysis capabilities of NISA:

* Plate/shell and solid elements available for modelling complex composite structures.

* Fiber directions may be specified globally.

* Composite layers may be isotropic, transversely isotropic, orthotropic, or anisotropic.

* Arbitrary layup with any number layers of multi-materials.

* Can model sandwich construction

* Same number of degrees of freedom as isotropic elements.

* Includes bending-extensional-twist-shear coupling even for flat plates.

* Smooth variation of lamination parameters (thickness, fiber angles, etc.)

* Includes interlaminar shear effects.

* Distribution of interlaminar shear stress is found.

* Contour plots of layer stresses and stress resultants.

* Survey plot of critical stress components in all layers.

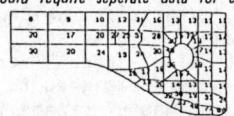

This cross member has multiple layers of continuous fiber-reinforced plastic. The element - independent angle input features of NISA-Composites allowed all fiber orientations to be specified in two compact data sets. Conventional programs would require seperate data for each element.

This stress survey plot identifies highly stressed regions of a composite structure and guides the engineer in making a meaningful search through the stress plots and printout.

NONLINEAR ANALYSIS

NISA can be used to solve a wide variety of geometrically nonlinear and materially nonlinear problems. The geometrically nonlinear capability includes stress-stiffening effect, large deflection, and nonlinear strain-displacement relationships. In the material nonlinearity, elastic-perfectly plastic material or elastic-plastic work hardening (isotropic work hardening) material can be analyzed. The Von-Mises, Tresca, Drucker-Prager and Mohr-Coulomb yield criterion can be used to model the yield surface. The nonlinear stress-strain data can be given in terms of explicit properties, as a piecewise linear stress-strain curve or a Ramburg-Osgood representation of the material.

The solution of nonlinear equations is obtained by an incremental and iterative scheme. User can choose modified Newton-Raphson or full Newton-Raphson iterative technique. The convergence is checked by displacement criterion and Euclidean norm using user defined tolerance.

Temperature dependent material properties can be used in the analysis. Properties can be given in the form of table of temperature vs. properties or in the form of polynomial function of the temperature. Anisotrpoic material properties can be used in the geometrically nonlinear analysis.

NISA has been very successful in predicting nonlinear response of a system in a complex situation. Very frequently NISA is used to predict energy absorbtion, residual stresses, permanent distortion and other quantities which require nonlinear analysis. The demonstration and verification problems are designed to show the user many useful capabilities like work hardening, loading-unloading type load history, snap through type response, static crash worthiness analysis, failure prediction, pre- and post-buckling analysis, progressive yielding and automatic load stepping.

NONLINEAR NISA HIGHLIGHTS

MATERIAL NONLINEARITY

* Von-Mises and other filed criteria

* Prandtl-Reuss potential function

* Work hardening
  1. No hardening
  2. Isotropic hardening

UNIAXIAL STRESS - STRAIN DATA

* Explicit properties: initial yield stress and work hardening parameter.

* Ramburg-Osgood curve: initial yield stress, secant modulus and hardening index.

* Multiple piecewise linear stress - strain data up to 6 pair points.

* Multiple two point curves at different temperatures.

GEOMETRIC NONLINEARITY

* Terms of Green Lagrangian Strain
  1. Linear terms
  2. Nonlinear terms

* Coordinate Update Key
  1. Each iteration
  2. Each load step

* Stress Stiffening

LOAD HISTORY

* Individually at each step

* 'N' increments with equal steps

* Automatic step size selection

ITERATIVE SCHEME

* Modified Newton - Raphson
* Full Newton - Raphson

CONVERGENCE SCHEMES

* Displacement, force and energy criteria

SELECTIVE OUTPUT

* Output at each load step or at every 'N' load steps.
* Output for selected elements and nodes
* Files for displacement and stress contours
* Nodal or Gauss point stress - strain history curve
* Displacement - force history curve at specified DOF
* Summary of highest gauss point and nodal stresses.

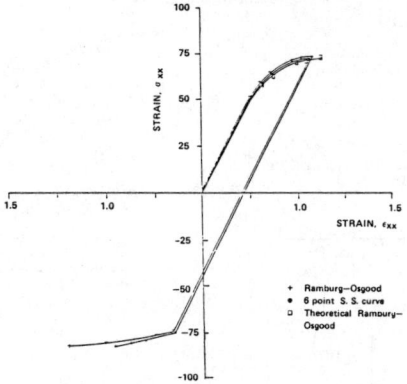

The uniaxial load cycling response of a bar. The elastic-plastic small displacement analysis is done using Ramburg-Osgood inelastic material model.

Stress–strain curve for Ramburg-Osgood type elastic–plastic material.

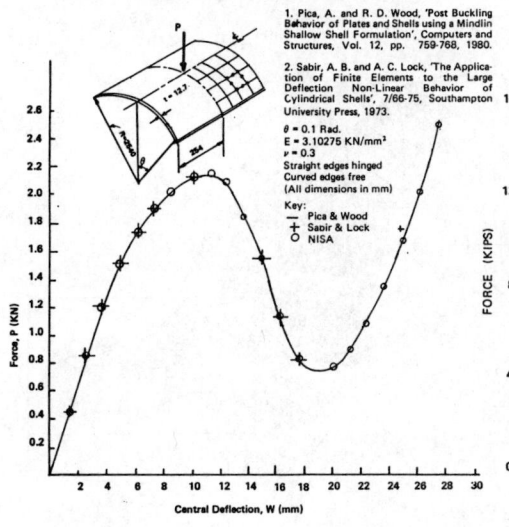

A hinged cylindrical shell with a point load at the crown is analyzed. The snap through behavior of this classical problem is accurately predicted using nonlinear strain - displacement relationship.

LARGE DEFLECTION OF A SHELL STRUCTURE
8-node supershell (semiloof) element of NISA used

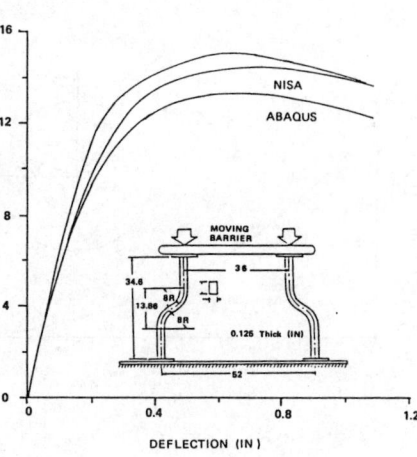

The static crush of a curved S-frame used in the vehicle impact test. A large displacement elastic-plastic analysis is performed using shell elements. Time consuming and expensive laboratory tests can be reduced by performing such analysis with NISA.

[1] Chi-Mou Ni, "Impact Response of Curved Box Beam - Columns with Large Global and Local Deformations", GMR Publ., GMR-1331, Warren, MI 1973.

Almen and Laszlo [1] have presented formulas which aid the deisgner in calculating the load deflection response of a Bevel washer (annular cone spring disk). NISA is used to calculate the load deflection curves for three types of simply supported annular spring disks. Shell elements are used for the nonlinear large deformation analysis.

[1] J. O. Alman and A. Laszlo, "The Uniform Section Disk Springs," Trans. of ASME, Research Papers, RP-58-10.

The front steering knuckle of a car is analyzed to predict the failure mode. A nonlinear large deformation elastic-plastic analysis of the steering arm is performed using 20 node solid elements. The effective plastic strain history helps in determining if the critical section is likely to crack or fracture completely.

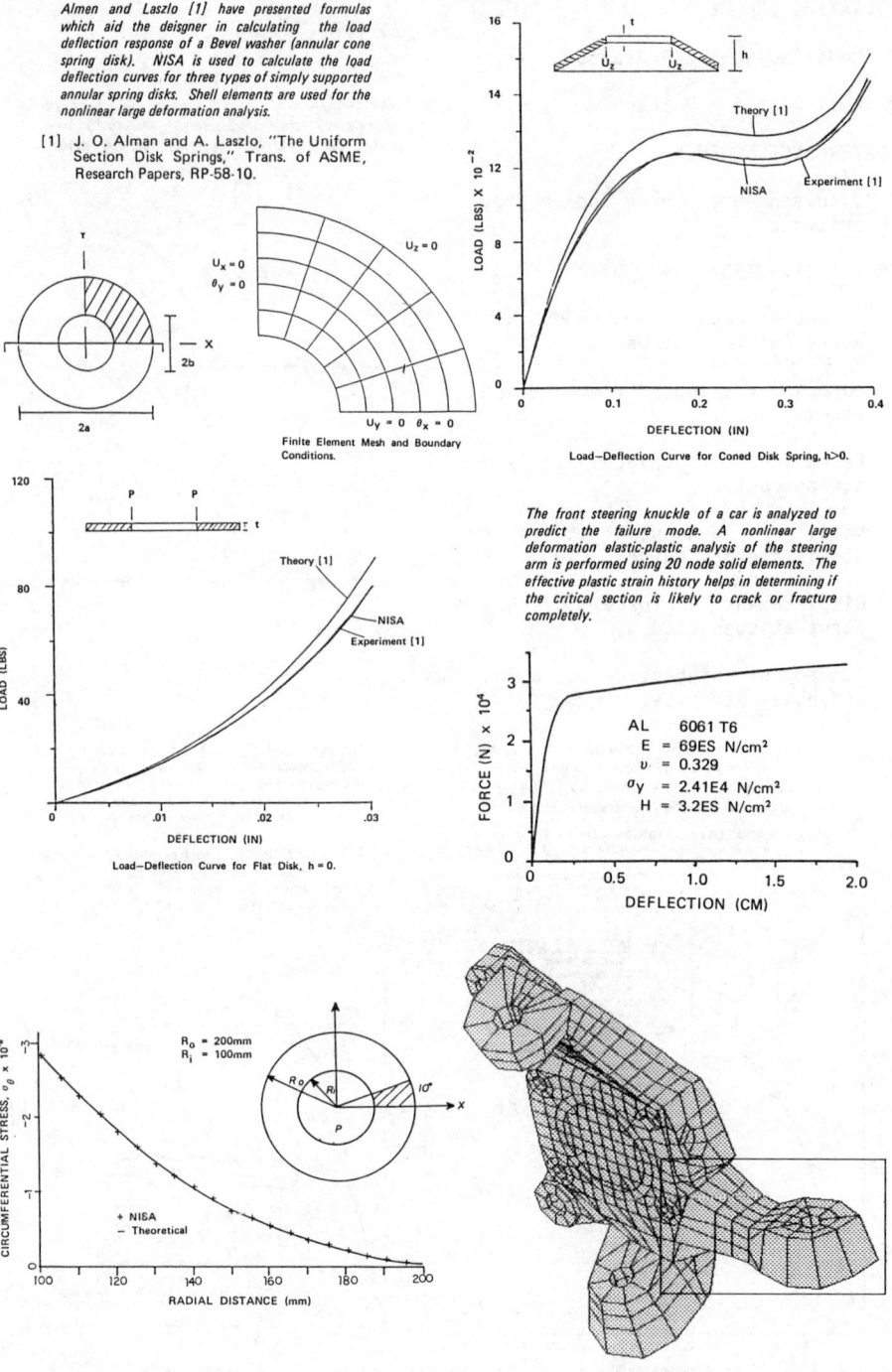

SYSTUS: AN INDUSTRIAL STRUCTURAL ANALYSIS SYSTEM AT THE SERVICE OF INDUSTRY

D. Halbronn and J. C. Chaumont

Framatome Computational Mechanics Center, B.P. 13, 71380 St. Marcel, France

FOREWORD

SYSTUS (the TITUS system) was designed and developed by engineers for engineers, in an industrial environment. It provides **the solution** - **the total solution** - to any structural analysis problem : heat transfer, mechanical analysis, linear or nonlinear, static or dynamic.

A complete, integrated, and user-friendly system, SYSTUS offers a wide range of pre- and post-processors, and extensive interactive graphics.

SYSTUS has a natural, free-format command language, very easily learned by the user, which controls the nature and sequence of each step of the processing.

Because it is an open system, SYSTUS allows smooth communication with other computer programs and easy integration into the industrial environment. Its modular form enables matching and following the user's developing needs.

SYSTUS is operational on numerous computers, from the supercomputers down to minicomputers and work-stations. It uses most graphics terminals currently in use (full color and monochrome).

SYSTUS : THE SOLUTION

Complete and customized versions of SYSTUS are installed in all branches of industry, as well as in universities and leading research centers. Its numerous uses in mechanical, civil, offshore, petrochemical, automotive, aerospace, and nuclear engineering, indicate the extremely wide range of possible applications.

SYSTUS capabilities make it possible to efficiently solve all thermal and mechanical structural analysis problems, in any industrial field.

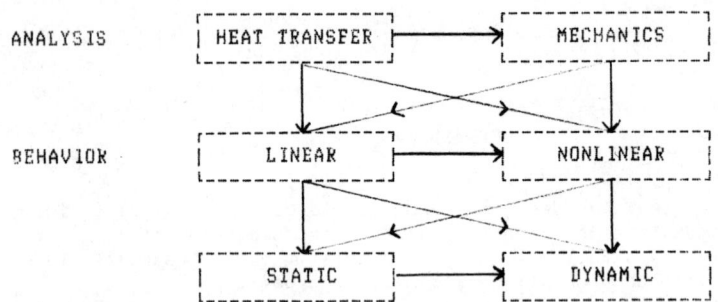

Heat Transfer

SYSTUS allows analyzing heat conduction problems (linear or nonlinear) under transient conditions as well as in a steady state. The calculation results (temperature maps) can then be automatically entered as loadings of a thermo-mechanical analysis.

The system also solve problems of forced convection and radiation.

This thermal option can also be used to deal, by analogy, with various other field analyses.

Static Linear

SYSTUS offers two methods of solving linear equations systems :

- the sparse matrix technique

- the banded matrix technique (with bandwidth optimizer)

Dynamic

SYSTUS includes many options for analyzing the dynamic behavior of structures, such as :

- calculation of direct response to sinusoidally varying loads, with or without damping

- response to arbitrary loads by direct integration (β Newmark method)

- fluid-structure coupling

- dynamic sensitivity analysis

- modal analysis. Depending on the nature of the problem to be analyzed or the type of results expected, the system allows choosing between several methods : Givens' method, inverse power method, or subspace method (Lanczos and iterative).

A response by modal superposition can then be obtained for the following loads :

- harmonic

- arbitrary (possibly taking account of confined nonlinearities)

- random (with calculation of transfer functions)

- spectral.

Nonlinear

With SYSTUS, several of the following nonlinearities may be combined in the same analysis :

- geometric nonlinearities : large displacements and strains, buckling and postbuckling, taking account of initial imperfections

- material nonlinearities : modulus of elasticity dependent on strain, thermal conductivity dependent on temperature

- plasticity and thermal plasticity, with various models of hardening (kinematic, isotropic, etc.)

- creep (with creep/plasticity interaction), visco-plasticity

- special mechanisms : gaps, stops, contacts with or without friction.

The solution of nonlinear equation systems, can be carried out through different iterative methods :

- the Newton-Raphson method ,

- the modified Newton-Raphson method, with or without acceleration (BFGS).

Note that SYSTUS incorporates the automatic Riks-Wempner incrementation algorithm.

Other Analyses

SYSTUS is particularly well adapted to solving problems of :

- fracture mechanics

- fatigue analysis

- welding simulation and determination of residual stresses (taking account of thermo-mechanical phenomena, but also metallurgical transformations, gas diffusion, etc.)

- electromagnetism (potential vector formulation, magnetic field formulation), electrostatics, electrokinetics.

SUPERELEMENTS

In the superelement (substructuring) technique, a structure that is too large to be analyzed all at once, is considered as an assembly of level n of superelements, each of which is an assembly at level n-1 of superelements of level n-2. The number of levels in such a tree-structure is practically unlimited.

Each superelement is processed as an independent problem, and the SYSTUS 'supervisor' automatically manages the data transfers between superelements.

This method enables :

- converting the analysis of a problem with a large number of degrees of freedom to a sequence of problems of a manageable size (therefore reducing the risks of errors and optimizing the use of your computer ressources)

- reducing the time necessary to analyse structures showing repetitive or mirror-image superelements

- analyzing alternative solutions of a problem where local modifications may be made, or following the evolution of a design as it is modified.

SYSTUS - THE TOTAL SOLUTION

SYSTUS effectively covers all stages of the analysis, whether for data input (model generation and validation), during the calculation itself, or the interpretation of results. This provides significant advantages during initial training and the everyday use.

Model Generation

Using SYSTUS, a finite element model corresponding closely to the real structure can be created :

- geometry (nodes, elements) : an extensive finite element library allows modeling all types of structures (elements with one, two, or three dimensions)

- materials : isotropic, orthotropic, anisotropic, incompressible, with constant or variable properties, and composite multilayered materials

- boundary conditions may be applied at nodes, or uniformly distributed on a rigid or elastic element (linear or nonlinear). They allow modeling of support conditions as well as simulation of special mechanisms (hinges, stops, gaps, contacts with or without friction, linked nodes, off-set beams,etc.)

- loadings (loads applied to the structure) : concentrated or distributed pressures, imposed displacements or temperatures.

Entry of the numerous parameters necessary for the finite element model is greatly eased by using automatic data input procedures, namely :

- automatic loadings (self-weight, centrifugal force, acceleration, etc.)

- automatic transfer to a structural analysis of temperature distributions resulting from a previously executed thermal analysis

- automatic mesh generation routines, to interactively generate one-, two-, and three-dimensional models, as well as models with translational or cyclic symmetry. These routines work, either from contours or macro-meshes interactively defined by SYSTUS or from the geometric model created by a CAD/CAM system.

- geometric transformations (rotations, symmetries, translations, etc.) used in the process of superelement modeling, including the automatic link generator (ALG). The user is also able to program his own rules for geometric transformation (the SYSTUS language contains a Fortran subset).

Model Validation

SYSTUS provides a wide range of data-checking pre-processors :

- data base content output (input data, generated data, default values, etc.)

- warning and error messages

- interactive display of the model on graphics terminals (monochrome or multicolored), or in batch mode on plotters with :

    . plotting of whole or part of the structure in any specified projection

    . zoom

    . section of a 3-D model

    . hidden line removal

    . nodes and element numbering

    . material numbering and identification

    . markers for boundary nodes and loaded elements

    . shrunken elements

    . solid modeling

    . graphics terminal multiwindowing.

Interpretation of Computed Results.

The engineer using the finite element method for structural analysis is too often left, at the end of the calculation process, with a large number of numerical results : displacements, temperatures, eigenvectors, etc., given for each node and/or element, and for each applied load, eigen mode, etc.

In SYSTUS there is a whole range of automatic routines, to process this information and present it in the most suitable form.

Numerical routines allow :

- combination and sorting of primary results

- their transformation (extrapolation, redistribution, merging)

- checking for compliance with the main contruction rules (ASME,RCC-M, NRC, CODAP codes, etc.), and a wide variety of graphic routines display :

- deformed geometry (in dynamic analysis, real-time animation of the modes of vibration)

- variation of results (response curves, time-history diagrams)

- contours (isotemperatures, isostresses, plastic zones, etc.)

- principal stresses.

All these routines leave more time to explore different alternatives of the model, and optimize its design.

SYSTUS - A USER-FRIENDLY SYSTEM

An industrial structural analysis system should not only be powerful, but also easy to use. SYSTUS' user-friendliness enables the engineer and designer to perform his task with the highest productivity.

Your Language

The user - most of all the occasional user - of a calculation system wants to solve his problem, without wasting hours in learning the intricacies of a computer program.

SYSTUS offers the language that enables the easiest man-machine communication. This language is natural : its convenient syntax, without constraints on layout (free-format) and its vocabulary, very similar to engineering terminology, are immediately understood and learned by the user. No specialized computer knowledge is required.

All computer tasks, such as data base management, restart and sequence of specific modules, are organized by the 'supervisor', which is the controller of the system.

In addition, SYSTUS is self-documented, and it is possible at any time to interactively access the on-line documentation.

Current developments are aimed at giving SYSTUS users the complete assistance of an integrated expert system : correction of errors in the data-input process, help in modeling, and choice of methods.

Interactive Graphics

Graphic display is fundamental in the structural analysis process, whether before the calculation (model creation and checking) or afterwards (it helps you make sense out of the information produced).

A continuous enhancement of the program, as well as the latest developments in graphics hardware give SYSTUS state-of-the-art interactive graphics (such as multiwindowing, real-time dynamic motion, menu-driven command, etc.)

At each stage of the analysis, SYSTUS' interactive graphics ensure cost-effective operation.

An Open System

Thanks to the open nature of the system achitecture, in-house programs can easily be integrated with SYSTUS. Several 'plugs' have been provided at pre- and post-processing levels, as well as at the calculation level (finite element formulation), to allow easy communication with in-house program developments.

The system can also communicate with other software (mesh generation systems, CAD systems, specific post-processors, etc.) by means of standard interface procedures.

SYSTUS IS TAYLORED TO FIT YOUR NEEDS, AND INTEGRATES WITH YOU YOUR INDUSTRIAL ENVIRONMENT

Complete SYSTUS versions are available on most of today's computers, from workstations running under the Unix operating system, to mainframes, and supercomputers. All graphics capabilities are built-in features of SYSTUS, and therefore independent of the device used (plotter, monochrome or color graphics terminal).

This means there is a SYSTUS version perfectly adapted to each user's hardware configuration.

In addition, thanks to SYSTUS' modular form, you can start with a smaller version meeting your initial requirements, then progressively add complementary modules, as your problems grow in size and complexity.

The finite element analysis system for which you are looking must be able to integrate with your own environment and thus enhance your actual investment.

Interfaces with most CAD systems, or solid modelers, are available. A unique connection with the EUCLID system has also been perfected in cooperation with Matra Datavision.

Furthermore, SYSTUS can interface with the main experimental dynamic analysis systems available today (enabling the solution of problems of mode identification, sensitivity analysis, modal synthesis, calculated and measured modes, etc.).

In short, SYSTUS provides the indispensable calculation link, perfectly integrated with the complete CAE package, which means higher productivity in your activity.

SUBJECT INDEX

Analysis
  mechanical, 195
Application program, 169
Artificial intelligence, 3, 57

BEM, 123
BEASY, 167
Boundary element method, 169

CAD, 123
CAD/CAM, 185
CAE, 185
Complex loading, 107
Construction history, 177
Crack propagation, 107
Creep, 177

Distribution
  of displacement, 93
  of moment , 93
Dynamic, 185
Dynamically-excited structures, 11

Electromagnetic, 185
Expert systems, 11, 27, 39
  in mechanical engineering, 57

Fatigue, 107, 185
Finite elements, 69, 107, 185
  method, 169, 177
Flexible couplings, 57
Fluid flow, 185

Geometric nonlinearity, 177

Heat transfer, 185, 195
Heuristics, 3

Interfaces, 3

Jacobian transformations, 69

Knowledge representation, 3
KYOKAI, 169

Legendre polynomials, 81
Linear/nonlinear, 185, 195

Material nonlinearity, 177
Mechanical analysis, 195
MEF/MOSAIC, 175
Mesh generation, 69
Microcomputer, 177
MODULEF, 173

Nonlinear/linear, 185, 195

Quality assessment, 69
Quasi-optimization routines, 11

Restart, 177

Shape
  distorted, 69
  optimization, 185
Shrinkage, 177
Simulation, 107
Static, 185
Structural optimization, 185
  strategies, 39
Structure
  dynamically-excited, 11
Supervisors, 3
SYSTUS, 195

Thermal shell, 81
TITUS, 195

RAYMOND H. FOGLER LIBRARY
DATE DUE

BOOKS ARE SUBJECT TO
RECALL AFTER TWO WEEKS

APR 0 7 1987